To SEA *and* BACK

Donald Is Too Late!

To SEA *and* BACK

THE HEROIC LIFE
OF THE ATLANTIC SALMON

Richard Shelton

Atlantic Books
London

For Frank Buckland,
Victorian Naturalist and Christian Gentleman

First published in hardback in Great Britain in 2009 by Atlantic Books,
an imprint of Grove Atlantic Ltd.

1 2 3 4 5 6 7 8 9

A CIP catalogue record for this book is available from the British Library.

ISBN: 978 1 84354 784 6

Printed in Great Britain by the MPG Books Group

Atlantic Books
An imprint of Grove Atlantic Ltd
Ormond House
26–27 Boswell Street
London
WC1N 3JZ

www.atlantic-books.co.uk

CONTENTS

Preface   1

The Turning Point   3
Home at Last   7
At Glamis   19
Field Mice for Tea   27
The Start of a Double Life   43
At Altries   53
Drinking Like a Fish   63
Living and Learning   69
To Know the Ocean Blue   77
To Sea with the Smolts   93
Fat is a Fishy Issue   105
Of Bugs and Brood Stock   111
The Hydrographer's Fish   125
By Ebrie's Bleak Banks   137
On Lancelets, Lampreys and Teleosts   149
Children of the Sun   167
Of Salmon and Sea Lords   177
The Real Meaning of Life   185
The Pilgrims   201

Some Words of Thanks   206
Index   207

LIST OF ILLUSTRATIONS

Integrated Illustrations

p. ii Donald Is Too Late! H. Cholmondeley-Pennell, *Fishing: Salmon and Trout*, 1895.

p. 5 The Aberdeenshire Dee. Queen Victoria, *Our Life in the Highlands*, 1868.

p. 8 Gaffs. H. Cholmondeley-Pennell, *Fishing: Salmon and Trout*, 1895.

p. 11 Salmon below a spawning ford. F. Whymper, *The Fisheries of the World: An Illustrative and Descriptive Record of The International Fisheries Exhibition 1883*, 1883.

p. 13 Cock and hen salmon. Courtesy of Robin Ade.

p. 14 Salmon ova. Frank Buckland, *Natural History of British Fishes: Their Structure, Economic Uses and Capture by Net and Rod*, 1880.

p. 16 Hen kelt. Courtesy of Robin Ade.

p. 21 The salmon in the stone. Courtesy of Patricia Shelton.

p. 25 The author's first salmon. Courtesy of Richard Shelton.

p. 27 Julia Pastrana. Courtesy of Richard Shelton.

p. 29 Brown and black rats. Leclerc de Buffon, *Histoire Naturelle Générale et Particuliére: Des Quadrupèdes*, 1799–1800.

p. 31 Sir Sidney Smith. © Classic Image/Alamy.

p. 32 Spine from *Curiosities of Natural History*. Frank Buckland, *Curiosities of Natural History*, 1893.

p. 34 Frank Buckland holding a cast salmon. Frank Buckland, *Notes and Jottings from Animal Life*, 1882.

p. 38 Spencer Walpole. © reserved; Collection of the National Portrait Gallery, London.

p. 45 The author and his brother Peter wildfowling. Courtesy of Richard Shelton.

p. 47 Newly hatched alevins. Frank Buckland, *Natural History of British Fishes: Their Structure, Economic Uses and Capture by Net and Rod*, 1880.

p. 49 Atlantic salmon parr. Courtesy of Robin Ade.

p. 54 Elsie Carstairs and her children. Courtesy of Richard Shelton.

p. 55 Cock pheasant. Lord Walsingham and Sir Ralph Payne-Gallwey, *Shooting: Field and Covert*, 1889.

p. 56 Archibald Ross. Courtesy of Richard Shelton.

p. 59 Jock Scott salmon fly. H. Cholmondeley-Pennell, *Fishing: Salmon and Trout*, 1895.

p. 61 Split cane rods. H. Cholmondeley-Pennell, *Fishing: Salmon and Trout*, 1895.

p. 66 Salmon smolts. Courtesy of Robin Ade.

p. 71 Ramon-y-Cajal. © Interfoto/Alamy.

p. 74 European lobster. F. Whymper, *The Fisheries of the World: An Illustrative and Descriptive Record of The International Fisheries Exhibition 1883*, 1883.

p. 79 Scientific laboratory. Sir Charles Wyville Thomson, *The Voyage of the Challenger*, 1877.

p. 80 HMS *Beagle*. Charles Darwin, *A Naturalist's Voyage Round the World*, 1890.

p. 82 Charles Wyville Thomson and companions. Time & Life Pictures/Getty Images.

p. 85 Recovering a dredge on HMS *Challenger*. Sir Charles Wyville Thomson, *The Voyage of the Challenger*, 1877.

p. 86 Dredging and water sampling apparatus. Sir Charles Wyville Thomson, *The Voyage of the Challenger*, 1877.

p. 87 Sea lily. Sir Charles Wyville Thomson, *The Voyage of the Challenger*, 1877.

p. 89 Captain George Nares. © 2003 Topham Picturepoint.

p. 90 HMS *Challenger*. Sir Charles Wyville Thomson, *The Voyage of the Challenger*, 1877.

p. 95 Cormorant and shag. © Ivy Press.

p. 96 Three-spined sticklebacks. F. Whymper, *The Fisheries of the World: An Illustrative and Descriptive Record of The International Fisheries Exhibition 1883*, 1883.

p. 98 Mayfly imago. H. Cholmondeley-Pennell, *Fishing: Salmon and Trout*, 1895.

p. 100 Post-smolt salmon. Courtesy of Richard Shelton.

p. 102 The author in Arctic waters. Courtesy of the Atlantic Salmon Trust.

p. 107 Fresh-run hen sea trout. Courtesy of Robin Ade.

p. 108 North Atlantic krill. Courtesy of Rod Sutterby.

p. 112 Snail. iStockphoto.

p. 115 Cab horse. © Bettmann/Corbis.

p. 117 An early salmon hatchery. F. Whymper, *The Fisheries of the World: An Illustrative and Descriptive Record of The International Fisheries Exhibition 1883*, 1883.

p. 119 Frank Buckland holding a tile. Courtesy of Richard Shelton.

p. 123 Frank Buckland's salmon hatching exhibit from *The Field*, 4 July, 1863. © The British Library Board, LD49.

p. 127 Atlantic salmon scale. Courtesy of Richard Shelton.

p. 129 Grilse. Courtesy of Robin Ade.

p. 130 Adult sandeel. © Ivy Press.

p. 133 *Johan Hjort*. Courtesy of the Atlantic Salmon Trust.

p. 135 Open cod end net. Courtesy of the Atlantic Salmon Trust.

p. 139 Craigdam Kirk. Courtesy of Richard Shelton.

p. 140 Young family and friends. Courtesy of Richard Shelton.

p. 142 Jar of worms. © Caroline Church.

p. 144 Immature sea trout. Courtesy of Robin Ade.

p. 146 Hen sea trout. Courtesy of Robin Ade.

p. 150 Lesser spotted dogfish. © Wildlife GmbH/Alamy.

p. 152 Lancelet. Paul Whitten/Science Photo Library.

p. 154 River Chess. Courtesy of Richard Shelton.

p. 156 Sea lamprey, lampern, Planer's lamprey and pride. William Houghton, *British Fresh-water Fishes*, 1895.

p. 159 Fossil teleost fish. Courtesy of Dr. Mark V. H. Wilson, Laboratory for Vertebrate Paleontology, University of Alberta.

p. 161 Lantern fish. Richard Ellis/Science Photo Library.

p. 164 Smelt. William Houghton, *British Fresh-water Fishes*, 1895.

p. 170 Drift netting. F. Whymper, *The Fisheries of the World: An Illustrative and Descriptive Record of The International Fisheries Exhibition 1883*, 1883.

p. 173 Male basking shark. Frank Buckland, *Natural History of British Fishes: Their Structure, Economic Uses, and Capture by Net and Rod*, 1880.

p. 175 North Sea cod. Frank Buckland, *Natural History of British Fishes: Their Structure, Economic Uses and Capture by Net and Rod*, 1880.

p. 179 Humpback whale. © Ivy Press.

p. 181 Scrimshaw. © Tony Arruza/Corbis

p. 182 Whalers and harpoons. F. Whymper, *The Fisheries of the World: An Illustrative and Descriptive Record of The International Fisheries Exhibition 1883*, 1883.

p. 184 Minke whale. © Ivy Press.

p. 186 Balmoral Castle. Queen Victoria, *Our Life in the Highlands*, 1868.

p. 188 Flask. © Caroline Church.

p. 190 Hen salmon. Courtesy of Robin Ade.

p. 192 Leaping grilse. © Emily Damstra.

p. 194 Arctic Charr. William Houghton, *British Fresh-water Fishes*, 1895.

p. 196 Salmon trout. William Houghton, *British Fresh-water Fishes*, 1895.

p. 199 Grey whale. © Caroline Church.

p. 203 The otter. Mary Evans Picture Library.

p. 204 Victoria Bridge. Getty Images.

Picture Section

1. The salmon in the stone. Courtesy of Patricia Shelton.

2. Frank Buckland. © The Royal College of Surgeons.

3. The cover of the *Curiosities of Natural History*. Frank Buckland, *Curiosities of Natural History*, 1893.

4. Buckland's fish museum. © The Royal College of Surgeons.

5. Cock salmon in breeding dress. Courtesy of Robin Ade.

6. Hen salmon in breeding dress. Courtesy of Robin Ade.

7. Salmon Parr. Sarah Bowdich, *The Fresh-Water Fishes of Great Britain*, 1828. © The British Library Board, L.R 404.C.S, plate xxix.

8. Seven Salmon. Sarah Bowdich, *The Fresh-Water Fishes of Great Britain*, 1828. © The British Library Board, L.R 404.C.S, plate xxxvii.

Preface

THE sea-run Atlantic salmon is an extraordinary fish. Large and classically proportioned, during its relatively short life it bestrides the two very different worlds of fresh and salt water. Its migrations far into northern seas to grow and subsequent return to its place of birth to breed are nothing less than heroic, and that return, with its dramatic leaps at falls, offers a wildlife spectacle no other fish can match. Its presence – or its absence – is a vital indicator of the relative health of our rivers and seas. Over the millennia, it has accrued totemic as well as culinary importance to the societies that caught it and for most of the last one it has enjoyed the conservation benefits of laws carefully drafted to protect it. Down the long years, the ancient statutes and their successors kept the great fish safe until, over little more than two centuries, first the rivers and estuaries and finally even the seas themselves, fell foul of industrial man's capacity to plunder the priceless resources he had inherited. For the first time in its long evolutionary history, the survival of the Atlantic salmon as a species can no longer be taken for granted. Many populations have been lost or put at risk, but the enduring power of this remarkable fish to capture the imagination of naturalists and sportsmen has so far proved a sure shield against its extinction.

Since 1967, this fish has had a charity, the Atlantic Salmon Trust, all to itself. It has been my privilege to act as the Trust's Research Director over a period when scientists much more able than I have revolutionized our understanding of this great yet resolutely enigmatic fish. As observation and experiment have unfolded new insights into the lives of salmon, so the results have appeared both individually in the scientific literature and combined with the outcomes of earlier work in a number of excellent books that provide a systematic summary of current knowledge. It is far too soon to make any attempt to add to their number but my editor, Angus MacKinnon, has persuaded me that it is not too soon to put together a few discursive reflections of my own. The result is not just a book about the salmon, although its life certainly provides the connecting thread. It is also, taking its cue perhaps from Izaak Walton's description of angling as 'the contemplative man's recreation', a meditation about a good deal else besides.

Richard Shelton

2009

THE TURNING POINT

MIDNIGHT and, for the last time that summer, the kindly sun skimmed just above the far horizon to resume its shallow climb into the Arctic sky. To the east lay the bleak slopes of the Lofoten Islands, the split bodies of cod drying on their racks above the foreshore. A couple of cables to seaward, the high dorsal fin of a male killer whale scythed through the gentle swell as, with the rest of its pod, it tore into a shoal of herring. For the cock salmon cruising at the surface with its schoolmates, the presence of the herring was a welcome diversion. Part way through an exceptional third summer of sea feeding, he would never be large enough to outpace a determined killer, but he was a fast and manoeuvrable swimmer, no longer worth chasing when more vulnerable fish were there for the taking. Now was the time to dive again to feast on the lantern fishes and krill that the subdued light of the 'simmer dim' had brought to within 200 feet of the surface. A string of tiny bubbles streamed from his gill covers as his swim bladder shed enough of its gases to speed his sounding. Soon he was among the countless hordes, snapping mercilessly at fishes that scientists call myctophids and that, though he was not to know it, were his distant relatives.

Back at the surface before his stomach was really full, somehow he seemed to be losing his enthusiasm for tearing into these beautiful but feeble little fishes. Indeed, as day followed day and the equinoctial gales ripped across the creaming surface of the grey Norwegian Sea, so he began to respond differently to what the sensory hairs surrounding the magnetite particles in his lateral line were telling the navigational circuits in his brain. More and more his swimming took him toward the south and, before he was aware of what was happening, he and his schoolmates were crossing the edge of the continental shelf and were back in the same North Sea they had left as sprat-sized smolts a sea lifetime ago. The fact was that so rich had been the Arctic feeding that changes in light levels had triggered the earliest stages of puberty and, with it, the sequence of navigational steps that would return him to the highland burn where his life had begun.

He did not starve as, with the Shetlands to starboard, the little company made its southerly passage. The last of the windblown insects driven offshore by the westerlies from the bogs of Caithness were a pleasant distraction and, every so often, a sandeel shoal would make a serious addition to the store of calories he would need to fuel his reproduction. A pod of bottle-nosed dolphins posed a brief danger, but a sudden burst of adrenalin-fired sweeps of his broad tail saw him clear and, of his gallant party, only a seal-scarred straggler fell to natural selection's pitiless reaper. On past the wide richness of the Moray Firth and the knuckle of the 'Costa Granite' and thence inshore to the great beach of Lunan Bay. How strongly the river scent drew him as, close inshore now, he turned north into the residual current that caresses the east coast of Scotland. How lucky he was that the summer drought was long past, a time when many a salmon forced to remain offshore ends its life in the jaws of a grey seal hunting among the rocky headlands or the leaders of the hardy netsmen's so-called 'fixed engines'. Only a heron stalking the shallows saw his bow wave as, in the gathering dusk, he slipped over a sand bar into the peat-tinged waters of the River North Esk with their familiar hint of home.

It was not until he was well above the weir at Logie that somehow his brain first became aware that the sequence of scents bathing his nares was not that of home after all. The remnants of the last of the

4

sandeels lay in his stomach, but his appetite had gone and they and his fat stores were all he had left to sustain him in the river until spawning time over twelve long months away. Now was not the time to go in panicky, energy-wasting search of home, but to lie quietly in the deep slow water under the bank until the next spate. By the time it arrived he had long lost the powder-blue back and quicksilver flanks of his sea-going livery in exchange for grey-green above and a softer gleam to his sides. Back over the sand bar, he stemmed again the southerly tide and, swimming steadily north past Stonehaven's humble Cowie, he rounded Girdle Ness and encountered a scent so enticing that he knew for certain that home lay in the cool waters of Aberdeenshire's Royal Dee. A tiny burn on the Abergeldie estate had been the place of his birth more than half a decade before, but it would be the best part of a year before he would see it again, a year moreover without food and with only the reserves he had accumulated at sea to sustain him.

The Aberdeenshire Dee at the Linn

The secret of his survival would lie in long periods of quiescence in deep water out of the main flow. For months on end during the summer, the flow in the burn would not even have covered his back. So it was that a succession of deep pools in the main stem of the river would be his waiting rooms and, only when the rains of autumn had swelled his natal burn, would he strive with his fellows for the opportunity to contribute his genes to the next generation of his family. Whether or not he and his like would live to see those climactic days would depend on many things, not least upon the outcome of the last of his encounters with the works of the greatest and most dangerous of his enemies, descendants of the naked apes whose ancestors had first left their African homeland in penny numbers little more than a million years before.

Home at Last

Less than a short January day had passed since the lordly cock salmon had exchanged the cooling sea for the biting chill of the river in winter. Now, temporarily secure in the holding water below the far bank, he and a lucky few of his fellow pilgrims could afford to rest out of the main current. There, as the river hissed and bubbled overhead, he would finally rid himself of the irritant lice he had brought with him from the sea and digest what remained of its bounty in his shrinking stomach. Here for a time was peace, short weeks of dozing when the gentlest movements of tail and fin were all that were necessary to keep station, automatic reflex responses he could safely leave to the built-in circuitry of his resting brain. It was the collapse of the high pressure system that had brightened the days and frozen the nights that first signalled the end of his reverie. As the air pressure dropped, so snow flurries became heavy falls and, with the thaw, driving sleet gave way to a downpour so fierce that even the ice in the high corries began to break away to load the rapidly rising river with that chilling mixture hardy fishers know as 'grue'.

Somehow the salmon endured the onslaught, pressing ever closer to the river bed and moving aside only to avoid the rumbling cobbles and smaller pebbles the wrathful current had dashed from their summer resting places. Imperceptibly, as the depression passed over and the glass in the fishing hut rose, the flow began to slacken, and the alerted salmon resumed its upstream journey. It was then that, as if in a dream, he saw the bright flash and felt the pulsation of one of the hatchet fishes he had last seen long ago in the Norwegian Sea. Two sweeps of his broad tail and the prize was his, but with it came a sharp prick and shortly after a strong tug and the unpleasant realization that, for the first time in his life, he was being led captive by powers he did not understand. Bewilderment gave way to panic and a frenzied dash across the river. Briefly, he lay quietly in the lee of a rocky slab. The uncomfortable sensation in his mouth remained but, as he nosed slowly out from behind the rock to make his way upstream, he had the feeling that he had regained the freedom to go where he wanted and had somehow escaped the strange force that had threatened to take that freedom away. It was the resistance he felt as he turned into the main current that reminded him that he was yet a captive. Blinded to all pain by the endorphins released by the adrenal hormones that now programmed his muscular form for 'fight or flight', the enraged fish shot to the surface, shaking his great head as he burst out into the soft light of an east coast February.

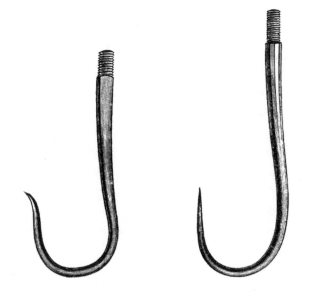

Gaffs for landing salmon – cruel relics of the past

Dash followed dash and long sulking runs, but still the fastening remained until at last the cramping lactic acid that had accumulated in his tired muscles stole the last of his strength and, turning on his side, he felt himself drawn helplessly into the knotless meshes of the ghillie's waiting net.

'Fit a gran' cock fush, Colonel, he most be a' o' twenty pun'. Dae ye think we should keep him?'

'No, Robbie, thirty years ago I would have said yes, but springers like him are all too rare these days and his genes are too valuable to take out of the river.'

The barbless hook fell away easily and, slipping the vanquished hero out of the net, the old ghillie held the gasping fish in the smooth current of the backwater below the fishing hut. The labouring gill covers slowed at last and the salmon righted himself; suddenly he was no longer there but secure under the far bank among the gnarled roots of an alder.

A day and a half later and the last of the lactic acid dispersed as the fish repaid the oxygen debt it had incurred during its long struggle. The levels of the hormones released during the fight were returning slowly to normal, and soon the salmon would be ready to resume the pilgrimage to the spawning fords for which his genes had programmed him. Ahead lay a succession of pools and bankside lies, places of quietness whose shelter could be won only by victory over the white water riffles that connected them. Spates were rare that spring, and by late June all too much of what remained of the flow was going to slake the thirst of the spray irrigators whose arching fountains were the life-blood of the potatoes and sugar beet that carpeted the haugh land with their moss green shaws. For weeks on end the riffles were impassable to all but the smallest fish but, even in the driest years, a maritime climate cannot be denied. Twice during the holiday months the sky over the Grampian foothills echoed to the rumbling crackle of summer storms and the rains that followed turned riffle to torrent and trickling fall to roaring cataract.

By now the great fish's sea-soft skin had thickened and toughened as another set of hormones prepared him for the final dash to the spawning fords and the risk of fatal abrasion it entailed. The final gateway to his calf country was a fall that excluded all but the fittest, fish like him being honed by natural selection to arrive in time to take full advantage of the earliest of the autumn water. He was drawn by the roar of the cataract toward the very sump of the fall, and the upthrust of the standing wave below it threw him bodily upward so that he landed among the bubbles and debris that now threatened to scour every scrap of moss from the rocks on either side. Powerful tail strokes drove him on and, for a few seconds, he made progress up the face of the fall. For a moment he was able to stem the flow, but the raw strength of the torrent first barred his way then swept him bodily downstream into the very tail of the pool. There, among a school of his doughty fellows, he recovered his poise and set his face once more toward the irresistible roar ahead. Again and again he leapt and swam, but still his ascent was denied. Then by chance he landed not in the tumbling core of the white water but in the smoother, more streamlined flow to one side. Here, in the boundary layer, he found to his surprise that he could swim faster upstream than the river could force him down; a final triumphant flick of his broad tail took him over the sill and into the pool above. Now at last he was home in the burn where, six long seasons ago, he first saw the soft light of a spring morning in the eastern highlands.

His were not the only eyes that lit on the strangely familiar surroundings. Smaller salmon that had entered the river in the late spring and early summer had caught him up. Most were grilse, fish that had enjoyed little more than a year of sea feeding, and most were cock fish like him. Like him also, their skins had thickened to a leathery toughness and had long ago lost the silvery gleam that had helped to hide them in the surface waters of the Atlantic. The patches of vermilion livery that relieved the greens and browns of his nuptial dress had been released from the fat store he had been living

A Victorian artist's impression of salmon gathering below a spawning ford

on since his long fast began many months before. Now the bright carotenoid pigments, which were derived from his marine prey, served to make him more threateningly conspicuous. A tartan tunic was not his only martial feature. As his skin thickened so did the size of his adipose dorsal fin to create a flag at the wrist of his tail that strikingly proclaimed his maleness. Most impressively of all, the finely wrought jaws that once plucked krill and small fishes from subarctic seas were now greatly elongated and so hooked into a gigantic kype that they could no longer fully close. Now at last he was ready to fight for the favours of the hen fish upstream, the first of which were twisting onto their sides to cut the depressions in the gravel, the redds, into which their precious eggs would shortly be shed.

As the hens worked, so minute quantities of ovarian fluid leaked from their vents, pheromone signals to the cock fish that the most important days of their lives were at hand. Gleaming softly with a hint of magenta, the hens had no need of oversized adipose dorsal fins or grossly projected jaws; neither were their skins decorated with the carotenoid pigment that had once reddened their flesh but now was redeposited in the rich yolk of the eggs. 'Cutting' a redd is something of a misnomer. 'Lifting' the fist-sized gravel with her tail flukes is what the hen fish really does when turned on her side. The upstroke of the tail reduces the pressure above the gravel so that it clears the river bed and the downstroke works with the flow of the river to push the gravel downstream. Every so often a hen tested the depth of her work by pressing the tip of her anal fin into the bottom of the redd. To the shadowing cock salmon, already lifted by her beguiling scent to a high state of arousal, this finny probing could so easily tip over into the spawning act itself and he responded by drawing close alongside. Shaken from nose to tail by waves of muscular contraction, his vibrating body drove shuddering pulses of infra-sound into the short space now separating the turgid bodies of the great fish. But it was not yet time and his watery foreplay was not matched by similar contractions by the hen.

Sea-run cock and hen salmon at spawning time

Temporarily distracted by his rough wooing, he did not at first notice the sidling approach of another cock fish almost as formidable as himself, but the instant he did so, he charged his rival with pitiless ferocity, his great kyped mouth wide open in slashing assault. Only the victim's toughened skin saved him from a life-threatening wound as he made good his escape among the roots of a bankside alder. Lesser fish, mere grilse that had enjoyed little more than a year of sea feeding, also challenged but were as easily intimidated into downstream retreat. Curiously, the great cock fish's real rivals made no attempt to challenge him. How could they, for most were but six inches long and weighed but a couple of ounces to his twenty pounds? Their strength lay not in their size but in their numbers and the concentrated potency of the sperm now swelling their tightening bellies. Good early feeding in the river and an innate tendency to become sexually mature while still at the parr stage had given these little fishes a first opportunity to pass on their genes without running the dread gauntlet of predation at sea. Like the sea-run males, the mature parr had also caught the ovarian scent and were now crowding round the vent of the cutting hen. Now and again her mighty paramour snapped at the crowd of upstarts and one of them, bolder than the rest, paid the price in a terrible wound, one that would become mortally infected as his sex hormone-weakened immune system was overcome by the ubiquitous bacteria infesting the spawning ford.

Another deep probe by the cutting hen and at last the cock's close shuddering brought both to open-mouthed orgasm. As the stream of apricot-tinted eggs poured from her vent into the very depths of the redd, so his spurting milt clouded the water above

Salmon ova at
the eyed stage

them. Some of his spermatozoa would find their mark in the micropyles of the eggs, but his were not the only male gametes to achieve fertilization. Deep in the redd, where nearly half the ova jostled, the parrs' concentrated ejaculate ensured that few went unfertilized. Not all the eggs found safe haven at the bottom of the redd. A few swirled about the rim of the depression, and some were swept out of it altogether where hungry parr, some already spent, snapped them up and in so doing helped replenish the energy reserves their frenzied lovemaking had drained from them. The best of them would recover sufficiently to join the smolt run the following spring; the remainder were fated to spend another hazardous year in the river, some to die of post-spawning infections, some to die in the jaws of predators, a few to spawn again as parr and fewer still to achieve smolthood and the dangerous freedom of the seas.

In the meantime, a succession of pulsing strokes of the hen's broad tail as she moved upstream covered the eggs in a pocket where most would sleep safely until the spring, the magic of embryonic development meanwhile transforming them into another generation of Atlantic salmon. So the gallant hen continued until her last pocket was completed. It was now time for her to leave, to let the tumbling stream carry her exhausted body out of the shallow burn and into the deeper security of the river. There, under the bank of an upper pool, her wasted muscles were just able to stem the flow. It was a time not of real rest but at least of respite, a time for her immune system to recover a little of its former germ-killing power and so allow her abrasions to heal. A winter spate swept her down to the lower river where again she sought shelter under a friendly bank.

She was desperately thin by now, yet her lithe body had begun to silver. A 'well-mended kelt' at last, her body shone with an almost unnatural gleam, a chrome-plated brilliance that would have done credit to a fifties Cadillac, so different from the tight-skinned bar of silver that had left the sea all those months before. Came the spring and the first of the schools of smolts swept past her. Stirred by endocrine echoes of her own smolthood, she cautiously left her shelter and let the river's swirling melt water carry her to the head of tide, over a sand sill and into the salty familiarity of the North Sea where a group of baby-faced seals had gathered to make easy prey of her fellow kelts. That she survived this dread onslaught owed as much to chance as to the boom of the netsman's .270 that temporarily put her would-be killers to flight. Survive, though, she did to rebuild her broken body with the sea's bounty, one day to return to the burn for a final tryst with new lovers among the rumbling gravel and hissing foam.

Back at the redds, the big male had long since forgotten his first spawning as a mature parr four years before, and even the memory of his recent conquests as a sea-run giant were beginning to fade. Fertilized eggs from three hard-cutting hens carried his genes and

Mending hen kelt

now he was nearly spent. Challenged by a large, recently arrived grilse, he could not secure a fourth hen for himself. He was weak and slack-sided, and only his aggression remained to fuel futile clashes with new arrivals and, finally, desperate visits to other spawning sites in the upper river where, as in his home burn, most of the hens had left. Knots of mature parr swirled around the clean, overturned gravel that marked the completed redds, and angry sea-run males splashed and fought in pointless battle.

For the first time, the great fish was no longer the captain of his soul. Blocked arteries and wasted muscles had stolen his strength, and his once-tough skin was now so sorely scarred by bacterial and fungal infections that it was no longer able to prevent the fresh water outside from entering his bloodstream and putting further strain on his labouring heart. Yet the will to live remained as he feebly sought to regain the lower river, mend his poor body and grow strong again in the sea. It was not to be; he had barely left the burn before he lost the ability to swim upright. Two days later, in the shallow tail of a pool, his great gill covers opened and closed for the last time. He died as gallantly as he had lived, immortal through the genes he had left to the salmon generations to come; the increased productivity conferred on the waters around him by the phosphates and other nutrients released from his decaying corpse were his final gift to the river that gave him birth.

Once, a little farther to the south and during what we have come to know as the Dark Ages, another cock salmon had sought genetic immortality among the current-winnowed gravels of a highland burn. Failing in this, his life's great purpose, he lived on, not through the DNA of his descendants, but thanks to the cunning hands of a long-dead stonemason.

———•———

AT GLAMIS

I T had been a gruelling three weeks, heaving the chosen monolith from where the ice of some distant winter had split the great sandstone slab from the rock face up the glen. Levered forward for mile after sweating mile on birch logs fresh cut from the Grampian foothills, at last it lay still on the sward alongside the exhausted warriors whose wiry bodies, adorned with strange, grey-blue designs, sprawled at rest until the gathering cloud of midges drove them into the turf-roofed huts of the settlement in which they would spend the night. The midges were still there at daybreak, ready to resume the wavering high-pitched singing that accompanied the exquisite torture they inflicted on all, but which only the youngest could hear. Blown into flickering life by the first to rise, smoke from the fire the women had lit the night before to cook their men-folk's barley-flour bere bannocks held the dancing multitude uncertainly at bay, until a welcome breeze from the coast beyond the low hills to the east gathered the strength to drive the devil's fairies back to the heathery bog that had spawned them. A few, a very few, would die among the sparkling glue of the sundews' beckoning leaves but, the moment the wind dropped, many of the females would be back to stoke their reproductive fires on the blood of their unwilling hosts.

The sun was well up in the late August sky when the clumping steps of an unshod pony were heard on the path that led up from its tethering post overlooking the river. A flaxen net lay folded over its broad back, and at its head stood the tall figure of its owner. As the spates of the previous week died away, so the salmon had leaped and jostled their way into the home pool below the village. Now was the time to gather the crop while the river was still low enough to shoot the net and before the next rise in water led the fish up into the glen to the pools below their stony spawning fords. Pagan to a man, none had heard of the new fishermen's faith that had swept across the Roman world just two and a half centuries before. Yet, had Simon Peter been present on the bank that day, only the way in which the wading fishers held the straining net against the flow would have differed from how he and Andrew, his brother, used a similar net in the still waters of the Sea of Galilee.

The tall man nodded. It was the moment to bring the net round, so that the foot and head ropes at both ends could be brought ashore and pulled steadily up the steep bank. For those like the over-excited boys helping for the first time, hauling a sweep net turned out to be surprisingly difficult. Taking up the slack was easy enough but, once all four lines were 'bar tight', so great was the resistance offered by the water and the round pebbles that kept the foot rope in contact with the bed of the river that nothing seemed to happen for minutes on end. At last the net began to move, the efforts of the fishers spurred on by the plunging bodies that bulged its broad belly. Out of the water at last, only when all was dragged safely up and over the shingle bank that stood between the river and the alders that fringed it could the party straighten their backs and admire the catch. Then it was that an older, fine-featured man, grey with years and with the large, calloused hands of an artisan, stepped forward to cast a critical eye over the thrashing forms. He was not looking for the largest fish or even the brightest and freshest-run. His quest was for the best proportioned fish. He was not to know that the choice he was about to make on that August day long ago would last for one and a half millennia and more. Seizing the salmon by the well-defined wrist at the base of the spade-like tail, he pulled a short truncheon or 'nabby' from his shoulder bag and killed the fish with a single, expertly applied blow to the head. In life, the chosen fish's eyes had been directed strongly downward in futile anticipation of the deep diving behaviour that is the salmon's instinctive response to a threat from above and that persists even when there is no water in which to sound. As the muscular body quivered in death, so the now sightless eyes returned to their resting positions in the centre of each orbit. Leaving the fishers to sort the catch, the old man slipped the fish into his shoulder bag and strode purposefully back to where the sandstone slab lay on the grass. Revered as a symbol of wisdom across the northern Celtic world, the representative salmon was about to enter its posthumous glory.

Selecting a twig of charcoal from the embers of the women's cooking fire, the veteran artist traced the salmon's outline onto the stone, taking care to pull each retracted fin into a lifelike attitude as he did so. Working quickly, he lifted the stiffening body onto the grass so that he could also draw in the details of the head and the position of the lateral line, a canal embedded in the skin that houses part of the salmon's close-range acoustic detection array and also the particles of magnetite invested with nerve cells that are thought to form one of the sensory inputs that enable salmon to find their way

out and then home across thousands of miles of the open Atlantic. Only when he was completely satisfied that the proportions of the sketch were exactly those of the fish did he cast the charcoal aside and patiently begin the long task of incising the traced image onto the stone. After two days of wrist-aching endeavour, broken by time spent sharpening his iron chisel and warring with the returning midges, he could sit back and look critically at what he had achieved.

The salmon in the stone

Yes, it was all there and, so accurately had the ancient master laid out his work, that the salmon biologist looking at the symbol stone today can be certain that the slightly elongated look of the head and the enlarged adipose dorsal fin are those of a cock fish, maturing but not yet ready to spawn. Even the tiny accessory pelvic fins, lanceolate spurs of tissue at the base of each of the second set of paired fins that are thought to reduce turbulence when the fish is manoeuvring, can be felt with the fingertips. Such was a tiny part of the precious legacy of a people who left no written records but communicate cryptically with those who came after through the symbol stones that mark their holy places. History knows them as the Picts, a corruption of *picti*, the painted ones, the name that the Romans gave to the fierce tribesmen they built walls to keep out and whom they met and defeated on the field of Mons Graupius during Agricola's second-century invasion of northern Britain.

The monolith bearing the salmon stands where it always has done, among soft grass, in what is now the manse garden at Glamis in Angus. After the salmon was incised into the stone, images of a serpent and of a mirror were added before it was pulled upright and embedded into the ground in a pit carefully dug to receive it. It is possible that the three symbols together mark the site as belonging to a particular tribal group, much as the spray-painted graffiti in our cities do nowadays. Intriguingly, the mirror is thought to be a female symbol, a reminder perhaps that Pictish royalty descended through the distaff line. Centuries later, after Columba and wave after wave of Scots had invaded Pictland from their Irish homeland and given their name to the kingdom, other artists carved a Christian cross, complete with a Celtic ring of glory, in low relief on the reverse side. Just how completely this later generation had left its pagan past behind we cannot really know, but the image of the head of a hind on one side of the cross and of a boiling cauldron with legs sticking out of it on the other suggests that early in the tenth century the process of enlightenment still had some way to go.

What the story of the symbol stone at Glamis tells us is that salmon were of iconic importance to Scotland's inhabitants long before they called it by that name. The fish in the stone and other, similar work by the Picts is among the earliest tangible evidence for the high regard in which the salmon was held in Europe. Yet there seems little doubt that the special status of the fish has a far longer history, one that perhaps reaches back as far as the emergence some thirty thousand years ago of Cro-Magnon Man, in whose middens, along the banks of the Dordogne, salmon bones are frequently found. Certainly, by medieval times the rights to fish for salmon were so highly prized, both north and south of the Scottish border, that they were a royal prerogative variously granted to leading subjects by their sovereigns in exchange for money or favours. (In England, one of the less well-known clauses in the Magna Carta of 1215 concerned itself with the removal of obstructions to salmon fishing in the Thames and Medway and from all places other than the sea coast, while in Scotland there was no public right to fish for salmon, including in the sea, a stricture that remains in place even today.) Indeed, the increasingly complex legal frameworks controlling the various forms of salmon fishing in the two kingdoms would continue to diverge, even after the Act of Union in 1707, with the result that the sustainability of the salmon resource was rarely seriously threatened by any kind of fishery until the mid-twentieth century and the introduction of the monofilament nylon drift net and the pelagic long line. As it was, the greatest danger of all was to come, not from fishing but from the obstruction and pollution of rivers and estuaries that followed the Industrial Revolution.

The Scots are, of course, still proud of their salmon. Although there are undoubtedly fewer wild fish than there were in Pictish times, they can still be seen and caught in all of Scotland's principal river systems, and at any time of the year in the highlands an adult salmon or one of its progeny is never more than ten miles away. Part of the explanation for this happy outcome is that Scotland lies at the heart of the southern European range of the species. Its rivers tend to be cool, clean and well oxygenated, and the substrates that underlie their headwaters and often the entire length of many of them are of cobbles, gravels and rocks at least as hard as the manse stone itself. It is to these environmentally demanding conditions that the freshwater phases of the Atlantic salmon are exquisitely adapted.

Although wonderfully blessed in this regard, highland Scotland has no monopoly on the salmon-dominated rivers of Great Britain. Similar physical conditions can be found in water courses in northern and south-west England and over large parts of Wales. Furthermore, there are other, often quite large rivers like the Severn and Hampshire Avon that, though not dominated by salmon, offer good opportunities in their headwaters, and patchily elsewhere, for salmon to reproduce in large enough numbers to support substantial fisheries. The Thames, and in continental Europe the Rhine, of which the Thames was once a tributary, were both lavishly endowed in this way until, over a few short years, the obstructing and polluting effects of the Industrial Revolution inflicted choking extinction on salmon populations that had abounded there for ten millennia. It was a tragedy that was to be repeated across the developed world on both sides of the North Atlantic. Looking back at those times through the murky looking glass of hindsight, it seems inexplicable to twenty-first-century Man that such incomparable treasure should have been cast so lightly aside. Perhaps even harder to explain is that so few stood up for the salmon at the time that the harm was being done. Of those who did, none was a better-informed friend of the fish than Francis (Frank) Trevelyan Buckland, Inspector of Salmon Fisheries for England and Wales from 1867 until his death at the age of fifty-four in 1880. The story of this most Victorian of naturalists, whose understanding of salmon rivers was nevertheless second Elizabethan, is truly remarkable and it is not only about salmon.

Better late than never, the author's first salmon (in reality a grilse)
at Altries, Aberdeenshire Dee

Field Mice for Tea

Just as the 'long' eighteenth century is held to have lasted from 1680 to 1820, so the Victorian age seemed to extend well beyond 1901 when the old Queen died at Osborne House in the arms of her grandson, Kaiser Wilhelm II. In the homes of many of her subjects, the atmosphere of her reign lived on for decades. In the forties, the ancient *Polystichum* fern in the middle of my grandmother's dining room table, the Zulu war club in the umbrella stand and the menacing horns of a long-dead highland bull – inspiration for many a childhood nightmare – were among the more overt reminders of an age when Britannia ruled the waves and London was the capital of an empire that circled the globe with red. Bound copies of the *Badminton Magazine* enabled one to ask such modern questions as to whether grooms or mechanics made the best chauffeurs and, in the lavishly illustrated *Living Races of Mankind*, the fetchingly topless Woman of the Tonga Islands competed for space with the body of the hairy and prognathous Julia Pastrana, there presented as a living link with Man's early

The body of Julia Pastrana, a marvel of Victorian taxidermy

ancestors, a most unlikely hypothesis given that poor Julia was fluent in three languages and had a sweet singing voice. Most interesting of all to my brother Peter and me was a group of red bound books entitled *Curiosities of Natural History* by Frank Buckland, Surgeon to the Second Life Guards. I have the four little books in front of me now and, opening the first of them, it falls open at the chapter on rats that was my favourite holiday reading of over half a century ago.

As a medical man, Francis Trevelyan Buckland would have been well aware that, on rare occasions, both the black and brown rat can be carriers of unpleasant infections, but he did not let that knowledge get in the way of his appreciation of both the black or ship rat, *Rattus rattus* (L.), and its nowadays more common relative, the misleadingly named Norway or brown rat, *Rattus norvegicus* (Erxleben), as the most intelligent of rodents. Back in the 1860s, Buckland surmised that the Norway rat was so-called because of confusion with the lemming and perhaps he was right. Modern opinion is that the brown rat came from central Asia from whence it spread north of the Himalayas into Russia and then, via shipping, to Great Britain. Just as the grey squirrel, *Sciurus carolinensis* Gmelin, tends to supplant the red squirrel, *Sciurus vulgaris* L., where the two species overlap in deciduous woodland, so the brown rat has out-competed the black rat over most of its British range. Apparently, however, *R. rattus* is still to be found in some of the best-known gentlemen's clubs in London and on the Island of Lundy, which the brown rat has yet to discover. Living in the East Neuk of Fife, and being but a rare frequenter of London clubs, I have only once seen what I thought at the time to be a black rat but that was almost certainly the far from uncommon melanistic form of *R. norvegicus*.

Buckland's knowledge of rats – 'It is not generally known what good eating young rats are' – was derived partly from his personal experience of wild and tame specimens and from his contacts with what were then called the lower orders of London and Paris, including 'Mr. Gibbons, a most intelligent and civil rat-catcher

Brown and black rats, note the finer features of the black rat

residing in the Broadway, Westminster'. Hearing that rat skins were used by French furriers in the manufacture of gloves, he 'inquired in many glove shops in London for gloves of this description' and tested the practicality of using rat leather in this way by tanning several rat skins himself. Still unable to satisfy his curiosity, we read of him 'keeping a sharp look out' for the characteristic appearance of rat hair under the microscope in the gloves of his friends and encouraged by the story of a Glasgow lady's shoes: 'A lady in town has just now a pair of shoes, of exquisite workmanship, the upper parts being made of the skins of rats. The leather is exceedingly smooth, and as soft as the finest kid, and appears stout and firm. It took six skins to make the pair of shoes, as the back of the skin is the only part stout enough for use.'

No doubt exotic footwear is still to be seen on the streets of Glasgow, but I would be surprised to learn that the skins of rats still play a part in its manufacture or that, in the far south-west of England, leaders of fashion are still to be found promenading in rat-skin garb. For after considering the possibility of importing the skins from China, where, according to Buckland, rat meat was also esteemed at table, he quoted approvingly from an old newspaper:

'An ingenious individual of Liskeard, Cornwall, has for some time past been exhibiting himself in a dress composed from top to toe of rat-skins, which he has been collecting for three years and a half. The dress was made entirely by himself; it consists of hat, neckerchief, coat, waistcoat, trousers, tippet, gaiters, and shoes. The number of rats required to complete the suit was 670; and the individual, when thus dressed, appears exactly like one of the Esquimaux described in the travels of Parry and Ross. The tippet, or boa, is composed of the pieces of skin immediately round the tails of the rats, and is a very curious part of the dress, containing about 600 tails – and those none of the shortest.'

In his liking for rat flesh Buckland was by no means alone. Fresh meat was such a luxury aboard the ships of the sailing Navy that, on a long voyage, a dish of rats could be a welcome relief from the salt beef and weevil-ridden ship's biscuit that dominated the cuisine of both the crew's messes and the midshipmen's gunroom. Rat-based dishes were not unknown in the wardroom or at captains' tables either. Perhaps the most distinguished naval officer to esteem the eating qualities of the ship rat was Sir Sidney Smith, victor of Acre and the only British commander to defeat Bonaparte's army on land until Wellington's victory at Waterloo. Taking Smith back to England aboard the frigate

Carmen with dispatches announcing the final expulsion of the French army from Egypt, his shipmate Mr Midshipman Parsons tells us in his journal, 'Sir Sidney, among many peculiar eccentricities, asserted that rats, fed cleaner, and were better eating, than pigs or ducks; and agreeably to his wish, a dish of these beautiful vermin were [sic] caught daily with fish-hooks, well baited, in the provision hold, for the ship was infested with them, and served up at the captain's table; the sight of them alone, took off

The 'Swedish Knight', Captain Sir Sidney Smith R.N., victor of Acre and rat lover

the keen edge of my appetite.' I have eaten neither the ship nor the brown rat so have no direct way of knowing whether Buckland and Smith's liking for these rodents was soundly based. However, I have eaten the flesh of two other species of rodent, the North American grey squirrel and the European beaver (the latter hot smoked), and can confirm that both were excellent. Rabbits and hares, which are widely eaten, are not, of course, rodents but members of the related mammalian order Lagomorpha.

Turning the pages of *Curiosities of Natural History* was to be my first encounter with the person of Frank Buckland. He was a man with an insatiable appetite for gathering information about the world around him through his own direct observations and through seeking the company of working men whose callings gave them insights into the lives of animals denied to the conventionally educated. He would undoubtedly have agreed with Glubb Pasha (Lieutenant-General Sir John Bagot Glubb, founder of the Arab Legion) that 'certain illiterate old men, who have not cluttered up their minds with quantities of

My grandmother's copy
of Frank Buckland's
Curiosities of Natural History
(First Series)

irrelevant information derived from books, may sometimes achieve a profound wisdom and knowledge of human life'. Buckland's willingness to treat his fellow men as as much a part of the natural world as the creatures on which they depended for their livings was revolutionary in the second half of the nineteenth century and was to give him insights into the wise management of the earth's resources that, a century and a half later, all too few have acquired. In so doing, Buckland had unconsciously uncovered the truth that Man, especially civilized Man, is the most predatory and destructive of the great apes and that understanding his impact upon the natural world as both killer and despoiler of the habitats of the species whose space he shares is crucial to the future of our overcrowded planet. Especially do these truths apply to the fishing industry where modern developments in locating, catching and intensively cultivating fish have rarely been matched by the resolute application of science-based management. All too often, the consequences have been desperate for both the fish populations themselves and the habitats that support them. As one of England's earliest and most industrious Inspectors of Salmon Fisheries, Frank Buckland was one of the first to recognize that such abuse of what he saw as a God-given bounty could end only in the impoverishment of the abusers.

Frank Buckland was the eldest son of the Very Reverend William Buckland, DD, Dean of Westminster, but, at the time of Frank's

birth in 1826, a canon of Christ Church, Oxford and Professor of
Geology and Mineralogy in the University. According to one of
Frank's letters quoted by George Bompas, the brother-in-law who
was to become his biographer, 'I am told that soon after my birth, my
father and my godfather, the late Sir Charles Chantrey, weighed me
in the kitchen scales against a leg of mutton, and that I was heavier
than the joint provided for the family dinner that day.' It was a
curious start to a most curious life. His father William Buckland
was one of the leading sedimentary geologists of his time and a
friend of Mary Anning, the gentle discoverer of the ichthyosaur
and the fossil remains of many another saurian embedded in the
Dorsetshire lias. William's best-known professional contribution
was his *Reliquiae Deluvianiae*, which described mammoths and
a variety of extinct fauna, the past existence of which challenged
traditional belief and especially Bishop Usher's assertion that the
earth was only some six thousand years old. Buckland explained
that his discoveries were the remains of 'pre-Adamite' creatures
that walked the earth in the days before the Flood, but not all of
Dr Buckland's clerical colleagues were happy with his conclusion.
Indeed, one of the old friends from his schooldays at Winchester,
the Reverend Philip Shuttleworth, who later enjoyed preferment
to the generously endowed See of Chichester, expressed his own
pawky reaction in the couplet, 'Some doubts were expressed about
the Flood; Buckland arose and all was clear as mud.'

Clerical doubts notwithstanding, Buckland had a wide circle
of scientific acquaintances and was a particular friend of the great
anatomist Sir Richard Owen and the equally well-respected geologist
Sir Charles Lyell, whose ideas on the chronology of geological
processes reinforced those of his own. As his career developed, so
his circle of distinguished friends increased. By the time he was
appointed to that most august of 'Royal Peculiars', the Deanery
of Westminster, his dinner guests included the physicist Michael
Faraday, the astronomer Sir John Herschel, the engineer Isambard

Frank Buckland in his study holding a cast salmon from his
'Museum of Economic Fish Culture'

Kingdom Brunel and the zoologist Louis Agassiz. It was his son Frank's good fortune to be exposed to such intellectual giants at a critical stage in his own academic development. It is perhaps to these early influences that we can ascribe the clear thinking that was later to mark his contribution to the conservation of salmon. However, there was much more to the Dean at dinner than refined conversation. He was a great believer in finding things out for himself. He therefore had no hesitation, when offered the opportunity, in eating some of the desiccated heart of Louis XIV, nor, when visiting a continental cathedral in the company of Frank, in tasting what purported to be the blood of a martyr – 'dark spots on the pavement ever fresh and ineradicable'. Falling reverently to his knees and extending his tongue, the Dean was moved to exclaim, 'I can tell you what it is; it is bat's urine.' At table, this willingness to try anything could take bizarre forms, puppy, crocodile and garden snail all featuring at different times on Deanery menus. John Ruskin, a frequent guest, wrote that 'I have always regretted a day of unlucky engagement on which I missed a delicate toast of mice'. No doubt the mice were of the field variety trapped by Frank, who regularly toasted these 'small deer' for his tea during his schooldays at Winchester, rather than house mice with their strong flavour of acetamide, a nauseous scent also responsible for the characteristic odour of old socks.

Indeed, a determination to extend the national diet through promoting the consumption of rodents, donkeys and other foreign and native creatures, thereby relieving the hunger of the poor, a fondness for live animals of all kinds, six-legged kittens, possible mermaids, Siamese twins and giants as well as a willingness to make friends of all classes in pursuing his wide rural interests were perhaps the chief personal characteristics of Frank Buckland when, at the age of forty in 1867, he was appointed by the Home Office to be an Inspector of Salmon Fisheries. His fellow inspector was Spencer Walpole, the twenty-eight-year-old son of the Home Secretary of the same name and previously his father's private secretary. By modern standards, the post was well paid, Walpole's predecessor annually receiving some £900–£1,000. But his father was keen to avoid criticism for giving so large a place to his own son, so the salary, no doubt for each of the newly appointed inspectors, was reduced to £700.

The two men complemented one another surprisingly well. Buckland was an excellent field naturalist, but had an intuitive rather than a quantitatively analytical approach to problems. Contrary to popular opinion, intuition, which merely exploits the remarkable data-sifting qualities of the unconscious mind, can be as important in pointing the way to advances in science as it is in other cultural spheres. Of course, it is likely to be productive only if there is enough of the right data to be sifted. Fortunately for the salmon, Buckland had a greater practical knowledge of the species than almost anyone else at that time. He was as baffled as the modern sportsman by the adult salmon's willingness to take a lure in fresh water despite the fact that it does not feed at that time. However, he had watched the behaviour of individual salmon for hours on end and knew that their responsiveness to the fly varies markedly over quite short periods of time. In his own words, 'I cannot say why they rise [i.e., respond to lures]. It is generally before a change in the weather. I should be most obliged if you would note down barometer and thermometer when they do rise and when they do not. We may then get some idea. Depend on it, fish and animals [sic] have some feelings and susceptibilities which we men have not.'

It was not until 2006 that that most experienced of salmon anglers, Andrew Bett, published the results of analysing long series of just such records and showed beyond doubt that short-term increases in the responsiveness of salmon (the 'taking times' of the angler) are triggered by rises in barometric pressure. My former colleague and fellow wildfowler Stephen Keay, who has pioneered the rehabilitation of kelts (spawned-out salmon) in Scotland, has observed exactly the same phenomenon in his recovering captive charges at the Government's salmon research facility at Almondbank near Perth.

Buckland was also convinced that the sense of smell was of critical importance to a salmon in finding its way to its river of origin, a conclusion that was finally proved to be correct by Canadian fishery biologists working over seventy years later.

We are indebted to Buckland's second biographer, Dr Geoffrey Burgess, lately Chief Scientist at the Ministry of Agriculture, Fisheries and Food, for an insight into the classically educated Frank's problems with arithmetic.

'His ability to handle figures was atrocious and he generally preferred to ask somebody else to do his arithmetic for him. He once gave the number of eggs in the roe of a specimen of carp as 2,059,759, having obtained this figure by weighing the roe and counting the eggs in a known weight. 'I can guarantee the accuracy of the weighing,' he said, 'and also the calculations, which were made for me by Mr. Thomas, a professional accountant.' The final digit in these calculations carried the same importance for him as the first.'

We may also note that 'Whenever he went on an inspection which was likely to involve him in considerable expenditure, he would arrange his sovereigns in twists of paper, each containing ten coins.' In those days, when even the most generously rewarded of Government officials was expected to have a private income, the inspectors would doubtless have met their day-to-day expenses from their own resources. It might nowadays be thought that such a system, which depended entirely on trust, would encourage corruption. In fact, experience at the time showed the opposite to be the case. Private means were normally associated with those whose patrician background and education had inculcated a code of Christian values that left no place for dishonesty of any kind. Furthermore, the possession of private means conferred freedom from the fear of penalty or dismissal and therefore the power to speak frankly and openly to ministers and higher officials, even when the advice so given was politically distasteful. Only the dwindling ranks of hereditary peers in the House of Lords enjoy such important political privileges today, and our national life is the poorer for their loss of influence.

Spencer Walpole's gifts were very different from those of his fellow inspector. He had, in Burgess's words, 'a marvellous head for figures and an incisive mind able to penetrate the legal fog which drifted over every question concerning the river fisheries', prime examples of the very qualities that Winchester College ever sought to develop in its scholars. It used to be said in the heyday of the British Empire that the world was run by Etonians on the advice of Wykehamists, the

Spencer Walpole, Frank Buckland's fellow Inspector
of Salmon Fisheries

inference being that the Etonian painted the main features of the political picture with a broad, intuitive brush but left the fine detail to his more turgidly educated Wykehamist chum. However, in this instance the roles were reversed, for Buckland was the Wykehamist and Walpole the Etonian. The combination was nevertheless of the happiest thanks to the good natures of both men. As Burgess tells us,

'The duties of the inspectors were not always pleasant, for many people from mill owners to fishermen resented Government interference in what they regarded as their rights. Here Buckland's diplomatic and entertaining manner was of considerable value. 'If a close observer were asked to mention the chief quality which Mr. Buckland developed,' wrote Walpole, 'he would probably reply a capacity for managing men. He had the happiest way of conciliating opposition and of carrying even a hostile audience with him. It frequently occurred that the fishermen, at the many enquiries which his colleague and he held, looked in the first instance with suspicion on the inspectors. They never looked with suspicion on them when they went away. The ice of reserve was thawed by Mr. Buckland's genial manner; and the men who, for the first half hour, shrank from imparting information, in the next three hours vied with one another in contributing it.'

Walpole's character was equally attractive. In his daughter's words spoken long after his death, 'I sometimes wonder in how many wayside villages and fishing hamlets may still linger a memory of his genial courteous passing... There was a certain simplicity about him and a ready sympathy, very different from condescension, which was so natural to him that he made it seem natural to others.'

The tasks ahead of Buckland and Walpole would have daunted less determined men. Buckland in particular would have known that the salmon's Achilles heel lay in the fact that, during the crucial phases of its spawning and juvenile development, periods that for many individuals occupied over half of their lives, the fish were concentrated in what, compared with the open sea, was a very small quantity of water. Even before the degradation of so many rivers that followed the Industrial Revolution, this division of labour between the relatively unproductive fresh waters, where the young fish were born and reared, and the productive and much less restrictive subarctic mixing zones of the North Atlantic Ocean, where deep water rich in nutrients rises to the surface to fertilize the sea where most of them made over 95 per cent of their growth, restricted the total abundance of the species. That is why, until the advent of large-scale intensive rearing, the Atlantic salmon was a relatively uncommon and always expensive occupant of the fishmonger's slab. The de-oxygenating and toxic effects of the wastes generated by a growing and increasingly industrialized human population soon began to turn this uncommon fish into rather a rare and, in some of the worst-affected rivers, an extinct one.

Buckland was one of the first to realize that it was not necessary for the whole length of a river to be polluted for it to lose its salmon. A short de-oxygenated reach at the head of tide could deny access as effectively as the mill dams and the other physical obstructions higher up that were the new main enemies of the salmon and those employed to defend them. It was the inspectors' job to identify such impediments to the free passage and subsequent survival of salmon in each of the rivers they visited and then to submit a report recommending appropriate remedial action. The latter would often include suggestions for the design and construction of fish passes. Even today, this is something of a black art. Buckland's technique was to approach the problem from the point of view of the sensations likely to be felt by the salmon. Lacking access to such modern refinements as flow meters and closed-circuit television, Buckland thought nothing of stepping boldly into the river and wading fully clothed into the deep water below the obstruction, armed only with a bamboo pole marked in feet. No doubt the element of showmanship that accompanied such amphibious theatre was impressive to those he was trying to persuade to construct a fish pass. However, my own childhood experiences of wading in chalk streams welling up from deeply buried aquifers has convinced me that the pressure receptors in the human skin are highly sensitive to flow until the moment where they and the nerve endings supplying them are paralysed by cold, at which point all dermal sensitivity is lost, mercifully including that to pain.

But who are we to say that Buckland and Walpole's methods were not exactly what were required at the time? Contemporary observers certainly thought so, as in the following commentary on their inspection of the grievously degraded Tyne and Wear, which are only now recovering from nearly two centuries of abuse: 'They did not go to their business in kid-glove style, but in a plain, practical, British workmanlike fashion they set about their task, making an impression on the mind of the spectator to the effect that these were indeed the right men in the right place. Not only so, but as becometh cultured gentlemen, the inspectors were exceedingly courteous and affable to every member of the party who essayed a word to them.'

Rarely were they able to reverse the harm that had been done during Great Britain's Gaderene rush to industrialize, but they were able to slow its progress and to draw attention to it where it mattered, namely in the drawing rooms of the ruling class to which they both belonged. Perhaps the real legacy of those gallant men, and that of their opposite numbers north of the border, is that, unlike other industrialized nations such as France, Germany and the United States, Great Britain remains a stronghold for the Atlantic salmon, a stronghold moreover in which post-industrial improvements in water quality are enabling salmon to re-establish themselves in rivers that had not seen them since the Regency. We owe a greater debt to practical, Christian gentlemen like Buckland and Walpole than we know and perhaps, in a secular age very different from theirs, than we are prepared to admit.

THE START OF A DOUBLE LIFE

S ALMON were not naturally present in the River Chess, the small
and icy chalk stream that rises in the Chilterns and in which,
along with my two brothers and cousin, I became first a naturalist and
then a biologist. That salmon had once been there I had no doubt.
Had not the ancestors of the mysterious lampreys that had finally
convinced me that I should somehow make my great passion my
life's work, the trout we cooked over fires of hawthorn twigs and the
sticklebacks we took home to watch them build their nests on the bed
of our outdoor primitive aquarium, entered the thawing Chess from
the sea a mere ten millennia before? Yet pollution and obstruction of
the lower Thames, in which the crystal waters of the Chess lost their
sparkling innocence on their way to the Southern Bight of the North
Sea, had long since put paid to the fish allegedly so abhorred by the
London apprentices that their indentures sometimes specified that
salmon was to be served to them on no more than three days a week.
Indeed, it is probable that these so-called salmon were not the fresh-
run bars of silver to be found at Billingsgate but disgusting, spawned-
out kelts weakly gasping their last in the shallows from which their
rascally purveyors had drawn them.

As it was, in the summer of 1962 I handled my first live, adult
salmon in a plastic dustbin on the banks of the River North Esk,
which enters the North Sea just above Montrose in Angus. It
had been taken from the estuary not long before by sweep net, an
ancient method traditionally known in Scotland as 'net and coble'.

I was a more than usually idle undergraduate part way through a student placement at the Freshwater Fisheries Laboratory, Pitlochry, whose gifted analytical chemist, Alan Holden, was developing a means for detecting the presence of cyanide in the gills of poached salmon. Poachers who previously depended on nets, ripping with treble hooks and fork-like fish spears or 'leisters', had learned that it was possible to secure salmon by poisoning the river with Cymag, a proprietary chemical used to control rabbits and rats. The active principal is the hydrogen cyanide liberated when the Cymag comes into contact with water. Its effects in the river are devastating because it kills all life stages of the salmon and a good deal else besides. Rather surprisingly, salmon killed in this dreadful way are safe to eat and their grisly recent history cannot be detected by superficial examination. Holden's problem was that small quantities of the cyanide ion are naturally present in the gills of salmon, a fact of which a skilled defence lawyer could make much. What was required was a quantitative method that could demonstrate beyond doubt the much higher concentrations of cyanide typically found in a poisoned fish. Trials with live adult salmon were essential in the later stages of developing what appeared to be a promising technique. Such was the grim fate of the fish in the dustbin. It did not die in vain; it was not long before even the threat that a Pitlochry chemist would be called as an expert witness for the prosecution would be enough to induce even the most hardened of the poisoners to plead guilty on the basis of 'certificate evidence' alone. Partly for that reason, and increasingly also because of the ready availability of cheap farmed salmon, the poisoning of rivers has become a rare crime.

Not long after the death of the dustbin salmon, my younger brother Peter joined me for a camping holiday in the highlands. Fishing for trout to supplement our diet in the River Druie, a tributary of the Spey, we caught our first salmon parr. Fortunately for them, they were only about four inches long, having enjoyed but a single summer of growth. Had they been any bigger, I have a horrible feeling that

The author wildfowling in the Wash near Gedney Drove End with
his younger brother, Peter in the 1950s

we would have consigned them to the frying pan, a crime that could
have led to a joint appearance in the Sheriff Court alongside the
poisoners.

Both the adult fish in the dustbin and the wriggling parr that
nearly ended up in the frying pan are snapshots from a life cycle
that spans two quite different worlds. On the face of it, it makes
excellent biological sense for the salmon to reproduce in fresh water
and to shelter its small and vulnerable juveniles in a habitat where
food is short but there are relatively few predators, and to save the
rich but dangerous world of the sea for a period of explosive growth
to adulthood. By the time the young fish enter the sea as smolts – the
word is a reference to their silvery appearance and is derived from
the same root as 'molten' – they are big enough to out-swim many of
the predators that decimate the much smaller young of most purely
marine bony fishes. At the other end of the cycle, the large, sea-
fed adults are able to lay, fertilize and bury in the protective gravel

larger eggs than would ever have been possible had they never left the river. It all sounds rather sensible and, on the face of it, strong evidence for the kind of intelligent design believed in so strongly by Frank Buckland and many another prominent Victorian, including Benjamin – 'I am on the side of the angels' – Disraeli. However, all is not what it seems. It is time to step back a little and examine the many strands that comprise the life cycle of this most enigmatic of fishes and the reasons why, at least until the Industrial Revolution, it proved so successful.

The lives of most marine bony fishes start as small eggs floating freely in the plankton. They are acutely vulnerable then and even more so after they have hatched when, unless they are able to find enough of their microscopic food before the maternal legacy represented by their tiny yolk sacs is used up, they will die. The proportion that survives in any one year depends upon how well the date and place of their hatching coincides with peaks in the abundance of their prey or avoids those of their predators. Even the young of bony fishes that lay their usually larger eggs on the bed of the sea have still to run the gauntlet of the plankton. It is a life in three dimensions, a life so chancy that in fish like the haddock, *Melanogrammus aeglefinus* (L.), the level of survival from egg to first appearance of the young fish in the trawl may be as much as a hundred times larger or smaller than the year before. A generation of salmon has a more secure beginning. The eggs are buried in gravel by their mother and thereby protected from most predators in that clumped group of pockets, the redd. The main danger to the eggs at this time is the interruption of their respiration as a result of smothering by silt or, as my former colleague Alan Youngson has recently shown, the upwelling of groundwater containing little or no dissolved oxygen. By the standards of those of most bony fishes, salmon eggs are relatively large so that when the larval salmon, called alevins, hatch in the spring, each is supported by a substantial yolk sac. As this food supply is depleted, the young fish make their way into the upper layers of the gravel where various

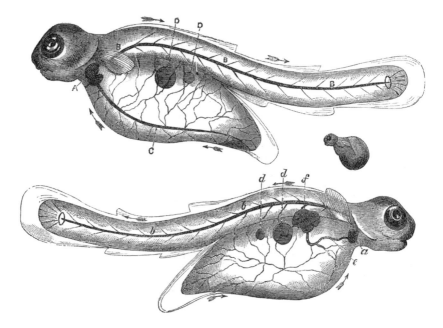

Newly hatched alevins of Atlantic salmon

predators lie in wait, including the earwig-like larvae of the larger species of stonefly, primitive but highly successful insects with a history of some three hundred million years. The bed of the stream onto which the young fish emerge in April or May is a much more dangerous place for soft-bodied creatures still encumbered with the remains of the yolk sac and too small to swim fast. They are still at risk from large, crawling invertebrates, especially dragonfly and beetle larvae, but greater dangers are posed by fast-moving mid-water predators, including trout, dippers, kingfishers and larger fish-eating birds.

At this, the most dangerous time of their lives, the fry have only passive defences on which to fall back. By emerging at night, they reduce losses to visual predators and, by doing so together, they reduce individual risk by saturating the finite appetites of their would-be killers. Temporary return to the gravel in response to the scent of predators or the shadows they throw offers some protection, but predatory mouths are only so big and one of the best defences of all is to outgrow them. Furthermore, swimming performance, and therefore the ability to escape mid-water predators, is related directly to size, so that by the time the young salmon are ready to migrate to sea, only large fish, piscivorous birds such as mergansers, goosanders and cormorants, and specialized mammals such as otters and seals pose a significant threat. Thus the more food the fish can obtain, the faster it can grow and the sooner it can shorten the list of animals capable of catching and eating it. The distinguished English fishery scientist David Cushing aptly called this process 'growing to avoid mortality'. If you are a young salmon, the essential first step in this life-saving activity is to win a feeding territory and achieve an independent life as a parr, the life stage that follows that of the emergent fry and lasts until the little fish is ready to migrate to sea, before the maternal legacy in its yolk sac runs out.

Yet the cool days of spring, when the little fish carries the remains of its packed lunch and is still able to enjoy the comforting security of the gravel whenever danger threatened, will soon seem an age away. In this grim period, threatening and being threatened by other parr vying to gain and hold a feeding station become as much a part of life as the dashes into mid-water to grab at drifting insects. There is nowhere near enough room for everyone and the route to survival lies in ruthless aggression to all competitors. Many of the little fish are already dead, the growing bodies of their killers their only memorial. As the days lengthen, however, so the intensity of the struggle slackens. The thumbprint-like 'parr markings' along the sides of the tiny salmon glow with that special intensity that signals local dominance. Indeed, life during the long summer days in the riffle in which the parr has claimed its feeding territory is not so bad. Overhead, swirls the full current of the burn, a conveyor belt for the drifting insects dashed at by the sharp-eyed parr whenever it instinctively feels that the size and proximity of the prey justify the risk of being seen by

Atlantic salmon parr

trout, goosander or otter. Most of the time, though, it sticks close to the gravel often in the shelter of a small boulder, the water flowing over its large pectoral fins – the homologues of our arms – outspread on either side of its head holding it within the boundary layer of slacker water near the bed. It is the relatively large size of these fins that enable the salmon parr to make the most of the opportunities offered by occupying feeding stations in a riffle, a habitat where a trout of similar size would, with its smaller pectoral fins, struggle to remain within that boundary layer.

If bright parr markings are the visual symbol of dominance within a young salmon's territory, the scent marking of stones is the olfactory one. Each unique mark says, 'Keep out or risk a fierce challenge', and together the marks help their owner to relocate its home range following displacement by a predator or spate. Interestingly, salmon are also aware of family as well as personal scents. They learn what their siblings smell like at the very start of their lives when crowded together as alevins in the gravel. For the rest of their juvenile lives, they appear more tolerant of siblings than of non-family members, and the results of tagging experiments hint at the fascinating possibility that some siblings remain loosely in touch even when at sea. Could it be that salmon have friends? Not really, it's just that they're not quite as nasty to their relatives as we sometimes are. Friendly to one another or not, salmon, like all living organisms, have persisted as a contemporary species because so far they have proved to be more successful at what they do, and therefore more capable of leaving viable offspring, than any of their competitors.

Salmon biologists are often asked what proportion of the new generation represented by the eggs deposited by a hen salmon survive to migrate to sea. It is a question to which there is no single answer. The fate of the young salmon varies greatly according to the suitability of the redd site itself and its proximity to food and shelter for the growing fish. Perhaps one of the advantages of the homing behaviour that is such a characteristic of the salmon and its relatives is that a fish that has survived to spawn is by definition the product of a place with a proven history of success. Whether this is the entire explanation of homing behaviour or not, it is certain that the hen salmon's choice of spawning site is critical to the eventual fate of her offspring, and that not all such choices are good. Even when the choice made is the best it can be, the level of competition between juvenile fish derived from different redds can make a large difference to the proportion that survives. There are also inherited differences between salmon families that can affect the 'fitness' of its members to cope with the demands of life in the river.

Perhaps the fairest way of summarizing the fate of a generation of young salmon from birth to migration to sea is to average the results of spawning in an entire tributary in which enough adults spawn to ensure that the eventual numbers of emigrant fish – they are called smolts at this stage – are set, not by the numbers of eggs laid, but by the amount of food and space available to the developing juveniles. Results obtained for a starting number of 5,000 eggs in the Girnock Burn in Aberdeenshire and the Burrishoole Fishery in Ireland by my former colleague David Hay, from whom I have learned so much about salmon, and the late Dr David Piggins are indicative. The 5,000 eggs survived well during their winter in the well-ventilated gravel, but brutal losses took place early in the first year as the newly emerged fry struggled to make an independent living in competition with their fellows. Conditions and survival improved markedly as the parr established themselves. Nevertheless, only fifty-two of the eggs initially laid gave rise to a viable emigrant or smolt.

One of the results of a selection regime even more ruthless than that which sieves out potential recruits to the Brigade of Ghurkas is that only the very best enter the next generation. There is another subtlety not obvious from such statistics but most important to our understanding of salmon populations. In well-stocked watercourses, the losses that take place in fresh water, especially during the first year, are partly dictated by the density of the young fish. The greater is the initial crowding, the greater are the losses and vice versa. The numerical effect of this so-called 'density-dependent' mortality is to reduce variation in the numbers of smolts which enter the sea – a variation that is reduced still further because each year's smolt run includes fish recruited from more than one year's spawning. As a result, the numbers of young salmon that enter the sea each year from well-managed rivers vary within remarkably narrow limits. Mark the contrast with the haddock, with its hundredfold variation in the strength of some successive years.

But we are getting ahead of ourselves. Before young salmon are able to survive at sea, each parr must first become a smolt. What happens then and later as a 'post-smolt' and pre-adult at sea generates most of the vagaries in the availability and behaviour of the returning salmon that renders its pursuit both so fascinating and so frustrating. To the angler, such caprice is part of his enjoyment, but to the ghillie it is a mischievous imp that can threaten his very living. It is time to visit one such ghillie at home, and we do so through the eyes of his granddaughter, Catherine Forrest, my dear wife Freda's late aunt.

———•———

AT ALTRIES

THE branch of the former Great North of Scotland Railway – a grandly named organisation whose locomotives were so small and feeble-looking that the rival drivers on the Highland Railway dismissed them as 'tin charlies' – which connected Aberdeen to Ballater was closed by Dr Beeching early in 1966. Many regretted its passing, but none did so more poignantly than Catherine Forrest, writing in that splendidly parochial newspaper the *Aberdeen Press and Journal*, known to all north-east folk as the 'P and J' but to one venerable English immigrant to the silver city as 'the two minutes' silence':

> Today the Deeside railway line is closed. For many people its memory will recall Royal comings and goings, stretching back to the beginning of the twentieth century and beyond. But for me, it was a more personal affair, the route to that enchanted world of childhood that no railway can ever take us back to, once we have grown up.
>
> The journey up from the south and the big railway station at Aberdeen had no thrills to compare with that familiar amble along the Dee to Milltimber; the nearest station to my grandparents' home. My mother always had her bicycle in the guard's van, for she loved to cycle those last few miles through the woods from Milltimber alone, shedding the years as she went. The children and the luggage followed her in a horse and trap (no buses in those

days), borrowed from the farmer and driven by Grandfather who would be waiting, in theory at any rate, on the platform when the train came in. In practice, my brother and I would sit there for what seemed hours, after the train had gone on its journey up the Dee. But, eventually, he would come, a tall, stooping figure, his brown face trimmed with a little, white pointed beard, and the whole lot topped by a tweed deerstalker stuck full of fish hooks. Grandfather was a gamekeeper and knew the meaning of silence. So did we, as we sat beside him in the cart, plodding easily along the red road through the woods. Then, at last, the trees opened up and there were the distant hills and the shining river and the

Elsie Carstairs (née Ross) and her children, Catherine and Andrew

sight that had been somewhere in
our minds all day, the house by the
waterside and my grandmother
waving to us from the door.

The hot scones on the girdle, the
raspberry jam, the stuffed pheasants
in their glass cases, the leather
game bags on the back of the door,
the rough walking sticks, the horsehair
chairs that scratched our legs, the ice-

Cock pheasant of the old
English strain favoured by
Archibald Ross

cold water we pumped from the well, the bere-meal porridge and
syrup, the paraffin lamps and the logs in the shed, the owls in the
wood and the moonlight shining on our beds as we fell asleep to
the sound of Grandfather's voice reading the newspapers aloud in
the room below.

It was a life that had changed very little, even from my mother's
childhood. They lived in a well-contained community of scattered
farms and homesteads, living by the seasons, utterly remote from
the life of the big city – not after all, so far away. Their only link with
it and the world beyond was the railway, unless, like Grandfather,
you were a prodigious walker and didn't mind an occasional route
march into Aberdeen to buy your morning paper. Even the walk
from Milltimber station was too much for my grandmother, laden
with parcels after a day's shopping in the town. She would take the
train instead to Peterculter, go by a short-cut through the fields to
the Dee and yell for Grandfather or one of the family to come and
row her back home across the river. A great-aunt of mine, choosing
the route for a surprise visit to the family, yelled on the bank all
afternoon in vain, and finally had to return to Aberdeen without
her tea. Sometimes we, too, arrived or departed via Peterculter.
Since there were so many hazards, like the children not walking
fast enough, my mother was always in a fluster about missing the
train. Once, as Grandfather was helping me into the boat, in my

Archibald Ross with his muzzle loader and pointers

black satin coat with blue forget-me-nots on the collar, I slipped through his out-stretched arms and plunged instead into the Dee.

As their family grew and departed, my grandparents became more adventurous and took quite long train journeys to visit them. Self-contained and taciturn as he was, after a lifetime as a gamekeeper, my grandfather nevertheless was as excited as a schoolboy on those train journeys and my grandmother was affronted by his enthusiastic efforts to engage everyone in the carriage in conversation, whether they liked it or not. Changing trains at junctions seemed to him a chancy business, and once, when he had lost sight of my grandmother on a crowded platform, he blew a blast on his dog whistle to summon her to his side. When my mother, all smiles and welcome, met them at their destination, they were, alas not speaking to each other.

But as time went by, they travelled less and we went more and more to visit them in their home on Deeside where, for nearly fifty years, they had seen their children and grandchildren come and go. I cannot remember the last time granny said to us, 'Haste ye back.' Those years are far away now and soon the railway, too, will be only a memory.

Archibald Ross, gamekeeper and my wife's great-grandfather, was born in 1849 at Marykirk in Kincardineshire, the part of north-east Scotland known as the Mearns and made famous by the gritty novels of Lewis Grassic Gibbon. Unlike many who follow this demanding calling, Archibald was not born to it but he had a country background to which was welded a better education than that of most of his contemporaries. His father, Andrew, was a schoolmaster, scion of a crofting family who ran what was then called an 'adventure' school at which his wife and cousin Mary taught knitting and sewing. It is important to emphasize that the adventure referred to had nothing whatever to do with the content of the curriculum, which would have concentrated on numeracy and literacy for both sexes, woodwork for the boys or 'loons', and household skills for the girls or 'quines'. The adventure was a reference to the financial risk taken by Andrew, in the days before state education was widespread, in setting himself up as a headmaster or 'dominie' of a small country school founded by himself. Most of the pupils would have left by the age of twelve and all would have been able to read, write and do simple arithmetic long before that time. The secret of such schools was to do a few things well, one of which was instilling the discipline without which, as so many state schools are finding today, even the most elementary skills cannot be taught.

Andrew died when his third son, Archibald, was only eleven years old. It was a dreadful tragedy at the time, but for Archibald it would one day be his passport to a magical life. Mary, his mother, remarried Alexander Adam, the head keeper at Fasque in Glen Dye, seat of the Gladstone family both then and now. For Archibald, who might otherwise have followed his father into the teaching profession, his stepfather's calling offered attractions against which slate, chalk and blackboard would have been utterly defenceless. What red-blooded boy, however academic his background, would want to spend his summer days ploughing through dusty textbooks when the rival attractions were setting tunnel traps for stoats and weasels and learning to cast a salmon fly? One is reminded of the reaction of a country loon at an Aberdeenshire Sunday school to the minister's account of the shadowy joys of the afterlife, a time when, allegedly, millions of years would be spent in the company of noisy choirs of cherubim and seraphim, rather scary six-limbed creatures that might well have been treated as vermin had they been found flapping about in the pheasant coverts of Fasque. 'Weel, weel,' the young worshipper commented, 'ah ken fit ye're sayin' meenister, bit ah wid jist like tae ging [go] roun' heaven wi' God, sheetin' rubbits.' So, no doubt, would have the young Archibald at the time he decided to become a keeper.

A decade or so after Archibald married Catherine Christie in 1871, and after experience as a ghillie and under keeper at Drumtochty

Castle in Kincardineshire, he secured the post of head keeper at Altries, near Maryculter, on the south bank of the lower River Dee. It was the most fortunate of placements. Within a short time, the Kinlochs of Altries had built a cottage overlooking the river for the young couple, a snug, substantially built little house at which they were to bring up thirteen children. They were on the threshold of the happiest days of their lives. The muzzleloader that Archibald had brought with him from Drumtochty was replaced by a brand new twelve-bore gun by Thomas Bland. It cost the laird six guineas, exactly six shillings more than six weeks of his head keeper's salary. It was adorned only by the maker's name and address and the proud legend 'The Keeper's Gun' along the top rib of its Damascus barrels, which were made from twisted bars of iron and steel heated and hammered together to provide tubes that combine tensile strength with adequate hardness, it was a plain but soundly made weapon which Archibald kept spotless. As to his salmon rods, they would certainly have been of greenheart rather than split cane and the chances are that they would have come from the same shop, Sharpe's in Aberdeen, at which he bought his silk lines, hooks and all his fly-dressing materials apart, that is, from the capes that he begged from cook every time an Altries cockerel took a one-way ticket to the kitchen.

Many of the salmon flies of today are rather dull affairs, plastic tubes dressed with feathers that slide down onto a treble hook. The best of them work perfectly well but, like so much modern sporting equipment, they are soulless productions. How different were the beautiful traditional patterns Archibald used to tie on winter evenings by the soft light of an oil lamp. Each was tied on a single hook or occasionally

The renowned Jock Scott salmon fly, a favourite of Archibald Ross

a double when, in high water, greater weight was required to get the fly down to where the fish were lying. His particular favourite was the Jock Scott, a pattern invented by Lord John Scott's water bailiff in the 1850s and widely regarded as the finest of all traditional salmon flies. Many steps and the feathers of several species of exotic bird went into its construction. To see a well-tied example is to gaze upon a true work of art. In the words of H. Cholmondeley-Pennell, doyen of Victorian angling writers and a former Inspector of Sea Fisheries, 'It would hardly be an exaggeration of language to say that this splendid specimen of artificial entomology has won an almost superstitious veneration among salmon anglers.'

Among Archibald's duties at the start of each fishing season was that of appointing temporary ghillies, partly to help him and partly to ensure that fish were not lost for lack of a competent attendant within shouting distance of the lucky angler. For the gentlemen and the occasional lady, the day would start at around nine o'clock at the fishing hut. Archibald and his team would have been there for some time, selecting flies according to the height and colour of the water and binding together the tapered joints of the majestic, double-handed greenheart rods with supple leather thongs. It was a method that was old-fashioned even in those days, but Archibald favoured it for the estate rods because there was no metal ferrule and socket to weaken the wood when a really good fish took hold and ran.

Greenheart is a tropical hardwood that can withstand repeated bending without taking a 'set'. It is also highly resistant to rotting and is much favoured for use as piling in the construction of docks and quaysides. Indeed, some rod-building firms were known to buy up old piling for turning into rod blanks. I have never used a greenheart rod but those who have assure me that the action is ponderous but reassuringly strong, rather like the black powder I sometimes use in my wildfowling guns. The snag with greenheart, compared even with split cane, is its great weight. As a result, the anglers of Archibald's time took their fishing at a gentler pace and did not flog the water relentlessly in the grim and joyless manner of some modern fishers. I suspect also that the weight of the rods was one of the reasons that there were fewer lady anglers at Altries in those days. No doubt those ladies who knew the form at Altries would have taken good care to bring one of the new split cane rods with them and with it outfish the male competition. Of course in those times of plenty, there was no question of 'catch and release' except for 'foul (pronounced fool) fush', that is to say inedible fish that were either close to spawning or were spawned-out kelts. The fresh salmon were either gaffed or tailed according to the preference of the ghillie and immediately knocked on the head with a nabby.

Early split cane salmon rod as used by lady anglers at Altries

The Aberdeenshire Dee in Archibald's time was a bigger river than it is today. It was just the same length, of course, but there was a lot more water in it then. Winters were longer and colder so there was more snow to augment the spring and summer flows with melt water. There were also few of the over-crowded plantations of conifers, whose roots draw water out of the ground that ought to find its way into the river, and whose drainage channels alternate flash flood with drought. Perhaps the most important difference of all was that Aberdeen and the villages and little towns strung along the Dee were all much smaller and so was the scale of the water abstraction required to dilute their drams and flush their lavatories. The greater wetted area made possible by this additional water increased the area available for the production of young salmon. It also eased the entry of adult fish returning to spawn so that fewer were forced to remain in coastal waters and at the mercy of seals, nets and dolphins during periods of low rainfall. The effect of all these differences between the river of Archibald's day and that of ours is that he and the 'big house' should have had far more salmon to fish over. However, at the time Archibald took up his appointment at Altries, there were no fewer than sixteen netting stations between Aberdeen and Banchory. The last of them was not closed by the Dee Salmon Fishing Improvement Association until the late 1880s, so it would have been some years before the Altries fishings approached their full potential and took their place as the principal sporting asset of the estate. Was there much difference between his salmon and those caught at Altries today? Records are scanty but what there are suggest that, at least in Archibald's latter years, the salmon were, on average, somewhat larger and included a greater proportion of early running fish.

Nearly a century after Archibald caught his last salmon, more and more evidence suggests that a good part of the explanation for any differences between his salmon and ours lies in what happens to the fish during their time at sea. Later we will join them there, but first we need to consider the uniqueness of water, the medium in which they spend their double lives and the challenges they face when they leave the river for the sea.

DRINKING LIKE A FISH

WATER is produced when two atoms of hydrogen combine with one of oxygen, something that happens billions of times every time we light a match or start the car. It occurs on earth as a gas, a liquid and a solid. It is a substance so universal that we tend to take it for granted, forgetting what a truly unusual compound it really is. Perhaps its strangest trick is to expand by 9 per cent when it freezes. It is the reason why ice is so buoyant. If it were not, the global circulation of water masses that moderate our climate could never have begun, polar seas would remain frozen from the bottom up throughout the year, the tropics would always be searingly hot and our own latitudes brutally cold for most of the time. Fortunately, our planet is spared this grim, life-shrivelling paralysis thanks to the structure and electrical private life of water molecules.

Physical chemists describe water molecules as 'polar' because of the positive and negative charges they carry. It is because of the inbuilt polarity of its molecules that water is such an excellent solvent, the best there is anywhere on earth. It is especially good at dissolving ionized and other particles that carry electrical charges. These include most of the salts and organic molecules essential to the complex chemical ballet that goes on in living cells. The ballet can proceed only if the constituent chemicals are gathered together in one place and separated from the surrounding medium by a membrane. It is only a partial separation. Water molecules are small enough to pass through a cell membrane, but those of many of the larger organic compounds such as proteins, carbohydrates and the various amino and nucleic acids are not. For a cell bathed in a dilute medium, there is therefore a greater concentration of water molecules outside the membrane than inside. The result is a tendency for water to flow into the cell, disrupt its chemistry and burst it.

This is a constant risk for a salmon in the river because fresh water is much more dilute than the body fluids that bathe its cells. Although its skin is reasonably waterproof, its gills and the membranes lining its mouth and pharynx are much less so and let in water. A little water is also drunk, probably mainly involuntarily when swallowing food. The fish protects its internal fluids from lethal dilution by pumping the excess water out through its kidneys, the long dark red structures at the top of the body cavity that most anglers misinterpret as congealed blood. Some salts are lost in the resulting urine, but these losses are made up by gains from the food and direct absorption through the gills. Sea water is much more concentrated than the body fluids of a salmon so that when it enters the sea its problem is to conserve water and exclude salts, the exact opposite of what it had to do in fresh water. It does this by actively drinking and excreting the excess salts across the membranes of specialized cells in its gills, gut and kidneys.

The miniature engineering that lies behind the salmon's ability to operate in two very different media is an extreme example of the function of cell membranes without which no contemporary life form could survive to live an independent life. Biologists are still not certain exactly how cell membranes are put together but most are agreed that, over much of their surface, molecules of substances called phospholipids (charged at one end but not at their long, non-polar tails) are arranged in a double layer with their fatty, uncharged ends opposite one another. Thus both surfaces of the membrane are electrically charged and this feature makes it difficult for other charged molecules to penetrate it in either direction. The tiny molecules of water are the exception, perhaps because of their small size or even the existence of narrow hydrophilic –literally water-loving – channels through the membrane. Proteins are also important components of cell membranes, both structurally and as the polar gateways that control the passage of charged particles in and out of cells. When the charged particles have to be transported against a greater concentration of charged particles on the other side of a membrane, as happens when a salmon in sea water excretes excess salt, chemical energy is required to power the process. The need to replace stores of chemical energy is one of the reasons why young salmon require access to a rich food supply high in calories from the moment they enter the sea in the spring.

For some parr, especially for those whose early lives were spent in high tributaries, the migration down to the main stem of the river begins the autumn before, but they still do not enter the sea until the spring, when the production of small mid-water crustaceans is increasing and the larvae of sandeels and those of many other marine fishes are hatching into the plankton in large numbers. Countless generations of natural selection have seen to it that there is usually a good match between the time of the year at which the smolts enter the sea and what biologists call the spring plankton bloom. However, just as in human affairs, there is no such thing as a perfect match and, in any case, the production of planktonic animals, especially of larval fishes, varies from year to year. The result is that in some years the smolts get off to a good start and in others they struggle to outgrow the ability of their early marine predators to catch and swallow them.

Leaving the secure world of the river involves a remarkable reversal of the fish's water budget, but it is only one of the changes that enable an aggressive territorial creature to confront the very different challenges of life in a small shoal of its former competitors in the sea. The patterns of spots and parr markings that had helped to hide it against a background of stones and gravel show up horribly against the blue of the ocean. One of the ways the young fish prepares for

Salmon at the smolt stage

life at sea is therefore to increase the amount of guanine in the skin. Guanine crystals are flat and silvery and so many are laid down during the transition to smolthood that eventually the fish's freshwater livery, including the parr marks, is hidden under a bright reflective coat. The guanine crystals are laid down in rows parallel to the light shining down from the surface and thereby act as mirrors that reflect the surrounding blue of the sea and so conceal the smolt from predators approaching from the side. The greenish-brown back of the parr is replaced in the smolt by blue so that the fish also appears sea blue to predators looking from above. Only from below is the blue theme abandoned. The fish's belly is a highly reflective white to reduce the risk of being outlined against the light shining down from above and attacked by a predator from below. Interestingly, the fins of smolts, especially the tail, contain a lot of black pigment that shows up rather prominently under water and may provide the fish with the species-specific visual cues it needs to hold its schools together.

Smolting also involves a subtle change in shape; the stubby shape of the parr is ideal for the sporadic bursts of acceleration that accompany its feeding behaviour in the river but the more fusiform shape of the smolt is better for sustained swimming over long distances. There is one other adaptation that does not affect the appearance of the young fish but which is absolutely vital to its successful reproduction as an adult. Well before it leaves the river, it needs to learn how to recognize the scent of home so that when eventually it spawns it does so at a site that maximizes the survival opportunities for its offspring. The capacity to learn is one of the reasons why the vertebrates, the group of so-called 'backboned' animals to which we and salmon belong, have enjoyed such success. But just what is it about the way that vertebrate nerves and muscles are organized that enables fish and men to profit from experience? It was a question that greatly interested my former Aberdeen colleague, Tommy Simpson.

LIVING AND LEARNING

DR Tommy Simpson, inventor, religious broadcaster and maker of some of the finest split cane fishing rods in Scotland, made his living as an endocrinologist, an expert on hormones, at the Torry Research Station in Aberdeen, an institute long since closed that sought better ways of processing the nation's fish supply. Its lasting memorial is the Torry kiln, a device for smoking fish without at the same time exposing an anxious public to the joys of botulism. How surprised the otherworldly men of Torry would have been to hear that injecting the dreadful bacterial toxin into the deeper layers of the skin is now the favourite pastime of leathery American film actresses. The fish that Tommy worked with were collected by the institute's own research vessel, a trim little side trawler called the *Sir William Hardy* after a pioneer of research into the processing of fish. Its reliability was not improved by the curious fact that it had at least four engines (I have a feeling that there might even have been a fifth of the same type to power the generator and winches) whose combined efforts drove the ship's propeller via an electrical transmission. On the funnel it bore the proud letters, DSIR, which stood for Department of Scientific and Industrial Research but which wags among the Aberdeen fishing community insisted stood for 'Don't sail if raining'.

Tommy Simpson had a special interest in developmental endo-crinology, the fancy name for studying the processes by which hormones are switched on and off as animals grow to maturity. He often used to say, not entirely jokingly, that if only he could discover a method for making cats behave like kittens for the whole of their lives or, for that matter, making sheep behave like lambs for the whole of theirs, he would make his fortune. Tommy was alluding to the fact that, for animals whose adult lives were based around specialized skills, the flexibility of behaviour made possible by an actively learning brain is disadvantageous, as well as being wasteful in the energy required to power the active transport of charged particles across cell membranes that lies behind all intense nervous activity, including our own conscious thoughts. We are lucky among mammals in that, in the words of Desmond Morris, 'Man's period of infant curiosity extends into adult life', but even our ability to learn tails off irreversibly as we grow older. It could well be that salmon are even luckier, but, before we consider why they might be, we must first address, along with Tommy, the more fundamental question of what it is about the nervous systems of backboned animals that lies behind their remarkable capacity to learn.

The great Portuguese biologist Ramon-y-Cajal discovered that it was possible to render individual nerve cells visible by staining slices of tissue with silver salts. It was a technique that magically turned what appeared at first sight to be a pale fibrous mush into a wonderful tracery in which tendril-like dendrites and gently swollen cell bodies join hands to form the on-board computers that co-ordinate the behaviour of all but the simplest of animals. Cajal himself once famously compared the microscopical appearance of the nervous systems of backboned animals with those of insects and their relatives. Whereas he described the former as like the workings of an elaborate clock, he reserved his greatest admiration for the latter, which, to him, were reminiscent of the finest Swiss watch. Perhaps without realizing it at the time, he had hit upon a fundamental difference in the ways in which the fine control of muscles is achieved in two of the most important divisions of the animal kingdom.

Ramon-y-Cajal in his laboratory in Santiago in 1906

In ourselves, the muscles over which we have voluntary control, like those I am using to write this sentence, are made up of thousands of individual fibres each of which at any one time is either fully relaxed or fully contracted. Fine movement in a limb depends upon the effects of nerve impulses from the brain and spinal cord varying the proportion of individual fibres in each muscle that are either fully contracted or fully expanded in this 'all or none' way. One only needs to watch a chameleon stalking a fly or a great craftsman at work on a masterpiece to be certain that the vertebrate system of motor control works well. Its great drawback is that to produce a gradual response in a muscle made up of thousands of all or none fibres requires that each of them needs its own nerve supply to switch it on or off. The consequence is that, for Jonny Wilkinson to convert a try or for Tiger Woods to sink a putt, requires the stimulation of many thousands of nerve fibres. Contrast these actions with those of a cockroach's leg as it runs across the floor or the claw of a lobster as it closes on its prey. The muscles of these creatures are based on bundles of what physiologists call myofibrils. Unlike our own motor units – in other words, each muscle fibre and its nerve supply – the response of a bundle of myofibrils is not all or none but depends on the pattern of nerve impulses it receives. As a result, a few tens of stimulatory and inhibitory nerve fibres are all that are required to provide a high standard of precise control. One result of this important difference in the ways its muscles are controlled is that the central nervous system of an insect can also fulfil its functions with a fraction of the number of nerve cells that a backboned animal would need to convert the information coming in from its sense organs into appropriate muscular reactions.

To Ramon-y-Cajal, the insect's nervous system appeared traceable and precise in its simplicity, whereas the mammal's seemed tangled and apparently chaotic, the relationship between structure and function not always being apparent from what he could see through the eyepiece of his Zeiss microscope. This was perhaps not surprising. The large number of nerve cells or neurones a backboned animal requires to control the contractions of its muscles also imposes demands on its central nervous system; the brain and spinal cord require even greater numbers of so-called interneurones to convert information coming in from the sense organs into an appropriate pattern of stimulation in the nerves controlling the muscles. The process of conversion requires that the interneurones must also have connections to one another. These may not always be fixed but modified and reinforced by particular patterns of stimulation. The result is what we see as learned behaviour, and it is perhaps no great wonder that it plays a much greater part in the lives of backboned animals, with their large numbers of interneurones, than it does in those of insects and their like. Thus the fuzzy apparent chaos of our central nervous tissue, itself a remote consequence of the inefficiency of our muscular control arrangements, enables us to speculate on the origins of the Universe, while the honey bee, with its Swiss watch of a nervous system, can learn little more than the direction from which its siblings last obtained a good feed of nectar.

I have never studied the behaviour of insects, but I have watched an awful lot of lobsters, prawns and crabs and marvelled at their sensitivity to light, sound, touch and the dissolved substances released from their food and one another. In locating their prey and their mates, in squabbling among themselves, in making their short breeding migrations and in securing their shelters, they give every appearance of purposeful activity. Looked at more closely, however, most of what they do can as easily be explained as reflex responses hard-wired into their rigid but nevertheless highly developed nervous systems. Only once was my faith in the essentially automatic nature of lobster behaviour shaken. A fellow biologist working on the other side of the Atlantic reported that he had conditioned an American lobster, *Homarus americanus* Milne Edwards, a closely related but slightly different species from our own, to associate a mild electric shock with another stimulus. As I remember, it was turning on and off a light. Conditioning an animal to associate one stimulus with another is a form of learning, so this news from across the pond was rather exciting to the inward-looking coterie of lobster-watching geeks to which, at that time, I belonged. I immediately tried to repeat the experiment, but after many unsuccessful trials I concluded that the European lobster, *Homarus gammus* (L.), had either failed to keep up with the literature or found the switching on and off of the light just as nasty as the mild electric shock. The lesson from all of this is that the ruthless forces of natural selection are quite capable of equipping the nervous systems of simple animals with appropriate responses to almost all of the situations they are likely to encounter. Pre-programmed from birth, animals so equipped need waste no time acquiring survival skills. This specification can be a winning formula where the environment and the response package that has evolved to cope with it are well matched.

European lobster, *Homarus gammus* (L.)

For those animals whose survival depends on learned knowledge and skills, however, the time spent learning is a time of increased danger. Not only may it lengthen the period required to reach reproductive adulthood, but it may also divert the energy resources of parents into care for their young, energy that might otherwise have fuelled greater numbers of offspring. Time spent learning tends, therefore, to be concentrated into a restricted period of the life cycle. You cannot teach old dogs new tricks because dogs acquire most of their life skills in their first year. In the same way, although the playful behaviour of lambs and kittens is flexible and to Tommy Simpson amusingly marketable, that of mature sheep and cats appears stereotyped and reflects their specialized lives as gregarious herbivores and visual predators respectively.

Apart from interacting with younger members of my own species as a father and occasional university teacher, and trying not to lose my patience when my dogs jump all over the furniture, I have spent rather little time watching the learning behaviour of mammals. My professional life has largely been concerned with the lives of fishes, most recently those of the salmon family. One of their great characteristics is that, at spawning time, they seek out the rivers, and even the parts of rivers, where they themselves were born and underwent their early development. It has been known for over half a century that returning salmon achieve this apparent miracle by remembering the smell of their freshwater calf country. More recently has come the realization that, as the adult fish make their way up a river system, their brains in some way retrace the memory of the sequence of scents they experienced when descending it as young fish.

My former colleagues John Armstrong and Nicky MacLeod are currently taking a more detailed look at the mechanisms that may lie behind what they call sequential olfactory imprinting in young salmon. It seems likely that the scents are recalled, not as a continuous succession like a video tape recording, but as a series of olfactory snapshots replayed in reverse order as the adult fish makes its way to its natal stream. The work is still in its early stages, but it could well be that these snapshots are laid down in the fish's memory during short periods when its brain is especially sensitive to minor adjustments in its circuitry. During their development, young salmon compete to occupy home ranges where the opportunities to feed and seek shelter from predators are best. As the fish grow and the seasons succeed one another, so their territorial needs change and they move from one home range to another. Some of the male parr even migrate within the river so that they can take part in spawning alongside the mighty sea-run adults. All of these migratory movements, including the eventual downstream journey to the sea, are accompanied by increases in the concentrations of thyroid hormones in the blood. Their effect is to facilitate greater levels of muscular activity. However, it seems that this hormonally driven 'turning up of the wick' does not stop there. It also stimulates the central nervous system, so that the salmon can learn and remember the sequence of local scents that enable the returning adults to find and spawn in the very part of the river in which they themselves were born. Thus, whereas in mammals the flexibility of the central nervous system declines inexorably with age, in a salmon it may be switched on and off at successive stages in its life according to its need to learn and to recall particular features of its surroundings.

We can but speculate about how a salmon feels during its learning phases. Perhaps the fairest way to describe these episodes is as periods of heightened awareness, but what constitutes 'awareness' in a fish's brain is as yet a total mystery. Of course, there is so far no scientific justification for drawing parallels with our own experience, but the fact remains that no sensation has a greater power to stimulate the recollection of past events than suddenly encountering a familiar scent. We should also remember that, setting aside the special adaptations of our circulatory systems to facilitate efficient air breathing by lungs, the bodies of fish and men share many characteristics. Our jaws and theirs started life as the gill arches of a common ancestor. Our ears open to the outside via modified gill slits. The paired fins of fishes are homologues of our limbs, and our teeth are not so different from scales. Indeed, women tend to develop dental caries during pregnancy in a way that closely parallels the erosion of scales in the run up-to the spawning in salmon. As Tommy Simpson frequently used to tell his colleagues, we are often closer to our fishy fellow vertebrates than we care to admit. If only we were close enough to share the salmon's instinctive ability to profit from the resources of the sea without doing them serious harm.

To Know the
Ocean Blue

THE bright salmon smolts whose little shoals tumble down our rivers with the first spates of May are true blue water sailors, off to seek their fortunes in the storm-lashed waters of the youngest of the world's great oceans. The Atlantic grows annually by the width of a man's thumb nail and now, after 120 million years, only the mighty Pacific exceeds it in size. Its northern waters are enriched by its own share of what oceanographers call the global conveyor belt of ocean currents. One of their local effects is to bring deep water, rich in nutrient salts like nitrates and phosphates, to the surface of the sea where, in the full glare of the sun, marine algae thrive, as do the wealth of zooplankton including crustaceans and young fishes that feed on them. One of the richest areas of upwelling is where cold Arctic and warm Atlantic waters meet along the north polar front. Here in the perpetual daylight of midsummer, plankton production proceeds with little interruption. It is to exploit these riches that salmon undertake feeding migrations measured in thousands of miles. They have been doing it for millions of years, but it is only since the voyage of HMS *Challenger* founded the science of oceanography in the late nineteenth century that Man began to understand the mechanisms that made sense of the salmon's life cycle.

There is perhaps no better way to pay tribute to the giants upon whose shoulders our gathering knowledge of the salmon's banqueting hall is being built, than to tread again *Challenger*'s rolling wooden decks. To do so is to travel back to a time when the Lords of the Admiralty were finally persuaded by men of science that, if Britannia hoped to go on ruling the waves, she would soon need to know a great deal more about what went on beneath them. We join the ship's company on a fine Sunday morning. Tucked away toward the end of that great celebration of sixteenth-century English, *The Book of Common Prayer*, is a section entitled *Forms of Prayer to be used at Sea*. The first and best known of them begins, 'O eternal Lord God who alone spreadest out the heavens, and rulest the raging of the sea; who hast compassed the waters with bounds until day and night come to an end: Be pleased to receive into thy Almighty and most gracious protection the persons of us thy servants, and the Fleet in which we serve'. It was then, and still is today, the prayer of the Royal Navy, and we can be certain that Captain George Nares RN would have read it to his ship's company every Sunday when the weather was calm enough for church to be rigged.

Taken at their face value, the words of the Naval Prayer interpret the cosmos in a way that long pre-dates the reformed Christianity of sixteenth-century England. They speak of an Old Testament universe specially created by an omnipotent deity capable of intervention in the actions of wind and wave to save the lives of those He may deem worthy of His favour. Is this what *Challenger*'s officers and ratings believed in the 1870s? We cannot know, only that they are far more likely to have believed it when desperately trying to shorten sail in the heavy seas of the Roaring Forties than when safely alongside in Portsmouth after their triumphant return from nearly four years spent in tortuous circumnavigation of the world. Even allowing for leisurely stopovers and dependence on sail for main propulsion, four years is a considerable time to spend showing the flag across a worldwide empire so vast that there was always some part enjoying the warmth of the sun. The explanation for the long absence lay in the real purpose of *Challenger*'s cruise. All but two of her seventeen guns had been put ashore before she left Sheerness on 7 December 1872 to make way for laboratories and storage cupboards, the accommodation of six scientists, their dredging and water-sampling gears and the miles of hemp and steel piano wire required to operate them. After 69,000 nautical miles the oceans of the world were a silent, myth-ridden immensity no longer. By the time the expedition's results had been written up and digested, a process that involved the compilation of fifty reports, the last of which was not published until 1895, so thoroughly had the science of oceanography been established that the world view represented by the Naval Prayer finally lost its battle with the even more wonderful truths being revealed by scientific endeavour.

With no serious challenge in home waters, the main task of Queen Victoria's Navy was to protect the trade routes and to act as an organic link between the mother country and 'Greater Britain', her increasing territories beyond the seas. To this end, naval officers

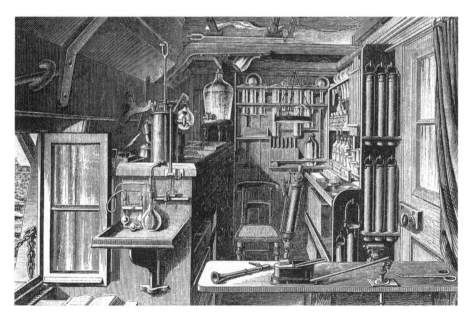

The well-equipped scientific laboratory of HMS *Challenger*

were encouraged to keep accurate journals of their encounters with distant lands and vessels and to record all manner of natural phenomena, from uncharted rocks and sea mounts to volcanic eruptions and sightings of real and supposed sea monsters. As the nineteenth century progressed, so official curiosity about the natural world grew and, with it, the practice of attaching civilian naturalists to small warships engaged in exploratory and imperial duties. Thus, Charles Darwin joined HMS *Beagle* in the December of 1831. His painstaking observations recorded in *The Voyage of the Beagle* formed many of the building bricks with which he was to construct his theory of evolution by natural selection and thereby achieve lasting recognition as the Isaac Newton of biological science.

So revolutionary was the outcome of the *Beagle* expedition that it is still regarded by many laymen as the most important research cruise of all time. However, from the point of view of the Lords of the Admiralty, the most lastingly useful products of the cruise were

HMS *Beagle* in the Straits of Magellan

not Darwin's later books *On the Origin of Species by means of Natural Selection* (published in 1859) and *The Descent of Man* (published in 1871) but the great contributions to the science of meteorology made in response to his *Beagle* experiences by Captain, later Admiral, Robert Fitzroy. Fitzroy's highly accurate patent barometer is still used by weather recording stations, and his name was commemorated in the shipping forecast when, in February 2002, it replaced Sea Area Finesterre. Nevertheless, whatever the Admiralty made of the results ultimately achieved by Darwin and Fitzroy, the idea that permanent good could arise from cruises, dedicated at least in part to scientific research, was well established in some quarters by the late 1860s. The Vice-President of the Royal Society of London at that time was William Carpenter, an expert not on salmon or the other fish and shellfish of commerce, but on such primitive zoological curiosities as sponges, foraminiferans (microscopic organisms, often with beautiful chambered shells) and echinoderms (the group that includes starfish and sea urchins). His great friend and fellow invertebrate enthusiast was Charles Wyville Thomson, Professor of Natural History at the University of Edinburgh. Wyville Thomson's eminent predecessor, Edward Forbes, had convinced himself that beyond the light and hydrostatic pressure levels prevailing at 300 fathoms, the world's oceans were devoid of life. Not everyone agreed. Having himself seen evidence of organic remains dredged from a Norwegian fjord over half a mile deep and having heard of the barnacles encrusting the frayed ends of a broken telegraph cable recovered from the bottom of the Mediterranean, Wyville Thomson no longer believed in Forbes's concept of an 'azoic' zone. He was determined to discover the truth for himself with the help of his friend, Carpenter, who, from his position of influence within the Royal Society, succeeded in persuading the Admiralty to make the steam frigate HMS *Lightning* available to Wyville Thomson for part of the wet and windy summer of 1868.

There, in the Faeroe-Shetland Channel, where a submarine ridge extending north-westwards from the continental slope now

Professor Charles Wyville Thomson (with gun) and companions
ashore during the *Challenger* expedition

carries his name, Wyville Thomson dredged living material from a carefully measured 600 fathoms and thereby dealt a further blow to Forbes's azoic hypothesis. He also discovered important differences in water temperature above and below 200 fathoms and thus the first evidence for the thermohaline circulation of the world's oceans. The scientific results of three further cruises, the first two by HMS *Porcupine* and the third by HMS *Shearwater*, convinced Wyville Thomson and Carpenter that a much larger expedition should be mounted to explore the sea beds of the world by the latest methods of sounding and dredging. The Navy was sympathetic but, even in the economically spacious days of the 1860s and 1870s, a certain amount of 'interest' was required to take even a minor unit away from its duties with the fleet, merely to indulge the curiosity of a handful of 'scientifics' and 'philos', as nineteenth-century seamen were inclined to call seafaring naturalists. (Today, they are often just referred to as 'effing scientists'.)

Here Wyville Thomson was doubly lucky. In 1870, his personal influence was increased by his appointment to the Edinburgh chair vacated by Edward Forbes. Secondly, the support he already enjoyed within the Council of the Royal Society through his friend and scientific collaborator, William Carpenter, was greatly strengthened by the sympathy of its President, Darwin's bulldog, the great Thomas Henry Huxley, naturalist aboard HMS *Rattlesnake* from 1846 to 1850. So it was that the Admiralty detached HMS *Challenger* from her front-line duties and made her available to Wyville Thomson for the first systematic investigation of the sea beds of the world. The objectives of the cruise, as agreed by the Circumnavigation Committee of the Royal Society, were extraordinarily ambitious, namely:

✳ To investigate the physical conditions of the deep sea in the great ocean basins (as far as the neighbourhood of the Great Southern Ice Barrier) in regard to depth, temperature, circulation, specific gravity, and penetration of light.

✳ To determine the chemical composition of sea water at various depths from the surface to the bottom, the organic matter in solution and the particles in suspension.

✳ To ascertain the physical and chemical character of deep-sea deposits and the sources of these deposits.

✳ To investigate the distribution of organic life at different depths and on the deep seafloor.

Even more extraordinary was that all four objectives were met, and with that achievement the realization among the most enthusiastic of the scientists that important marine discoveries are but rare interruptions to long periods of hard-working boredom, shorter ones of violent discomfort and occasional moments of sheer terror. That anyone should want to go to sea for anything other than personal gain would have been as incomprehensible to *Challenger*'s ship's company in 1872 as it is to the working seafarer of today. Certainly, any thought that Wyville Thomson and his five scientific colleagues may have entertained about the pleasures of oceanography as a diversion for gentlemen were dispelled the minute they left Sheerness and ran into a south-westerly gale of such ferocity that the ship had to put into Deal. Already downhearted after the loss of a young Royal Marine who had missed his footing on the gangplank and drowned in the filthy waters of Sheerness harbour, the officers and bluejackets were not best pleased when their six new shipmates, whose accommodation requirements had already disrupted their cramped quarters, elected to continue their journey to Portsmouth by train.

It was the worst possible start for an enterprise whose success would be utterly dependent on the ability of captain and chief scientist to weld their respective teams into a single determined organization that would go on functioning efficiently in the face of further tragedy, moments of great danger and, above all, week after

HMS *Challenger* recovering a dredge on her port side

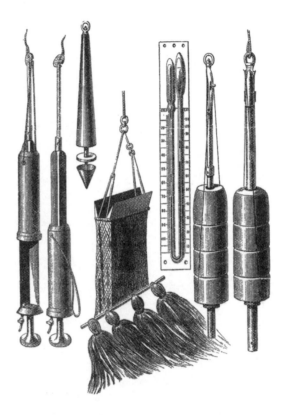

Dredging and water sampling apparatus

week of boring routine. Here Wyville Thomson's earlier seafaring experience stood him in good stead. He had the good sense to explain his scientific objectives to Captain Nares in detail and in terms that secured his total support. He had also been careful to choose colleagues not dissimilar in background to Nares's own officers, men in whom the concept of duty and personal honour had been instilled so early in their education that it had become second nature. Within weeks of leaving Portsmouth to begin the first of their ocean transects, the scientists had overcome their sea sickness and were sharing the hardships of the professional seamen: Wyville

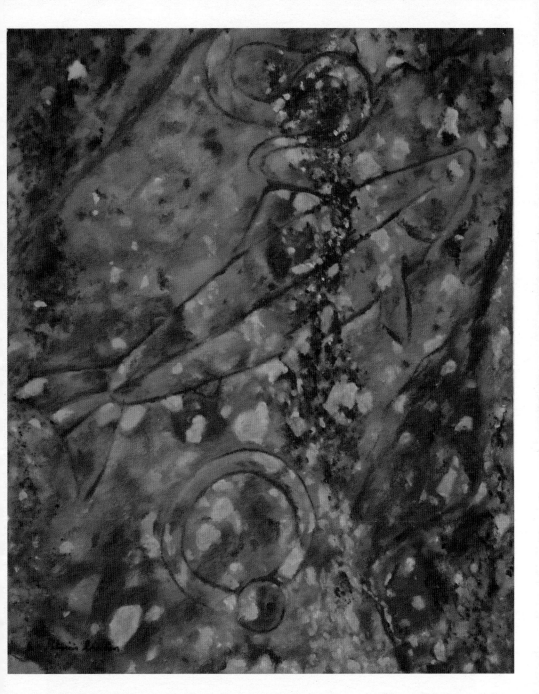

Patricia Shelton's painting of the Pictish symbol stone at Glamis with its patina of lichen and the incised image of a cock salmon.

The cover of my grandparents' copy of
Frank Buckland's *Curiosities of Natural History*,
first published in 1857.

Frank Buckland with his wading pole, note
the depth markings.

Frank Buckland's 'Museum of Economic Fish Culture'
at South Kensington.

Robin Ade's paintings of a cock salmon and hen salmon in breeding condition.

Early representations of a salmon parr and an adult salmon from
Sarah Bowdich's *The Fresh-Water Fishes of Great Britain*, 1828.

Thomson and Captain Nares found themselves in a long Atlantic swell at the head of a single band of brothers. Total success could not, from then on, be assumed but, given what the Articles of War – a stirring legacy bequeathed by the seventeenth century – refer to as 'The Good Providence of God', there was now a reasonable chance of achieving a fair measure of it.

Forbes's azoic zone was not the only Victorian misconception about the life of the deep ocean. Because he saw it as offering rather a constant environment, Charles Darwin believed it to be a place where the pace of evolution was probably relatively slow and therefore a likely home of 'living fossils' and 'missing links' in evolution. It was one of the objectives of the *Challenger* expedition's programme of deep dredging to seek out such interesting creatures. Apparent success was achieved early on when, soon after leaving Lisbon, they dredged a 'sea lily' from a depth of 1,000 fathoms (6,000 feet). Sea lilies are not, of course, plants at all. The world of science knows them as crinoids, close archaic relatives of starfish, sea urchins and sea cucumbers, members of the same group of phyla, one of which would, in due course, give rise to ourselves. Seen at the time as early fulfilment of one of the main aims of the cruise, and thus a good omen, finding the sea lily was to prove rather an isolated event. Far from being a near constant environment to which its inhabitants had become perfectly adapted for millions of years, the deep sea was revealed by the observations of *Challenger* and her modern successors to be a much more labile world, one subject to water movement, seasonal change and vulcanism. Yes, it does contain archaic organisms, like crinoids, lamp shells or brachiopods, six-gilled sharks and hagfish, but *Challenger*'s scientists were to see more echoes of the past in the egg-laying mammals of Australia, the platypus and echidna,

Sea lily dredged by HMS *Challenger*.

and the worm-like *Peripatus*, an arthropod belonging to the same group as crustaceans, insects and spiders, which they encountered under a rotting log in New Zealand, than ever they were to see in the samples they dredged from the deep.

Safe arrival in Melbourne marked the halfway point of *Challenger*'s odyssey. Astern lay the bodies of more shipmates, four crossings of the Atlantic, discovery of the mid-Atlantic ridge and the near loss of the ship among the gigantic seas and icebergs of the Southern Ocean. As revealed by the rating diarist, Joseph Matkin, life for the crew was grim indeed and grimmest of all when fog, ice and storm force winds offered the only respite from day after repetitive day of sounding and dredging. With the ship lying to under bare poles and with visibility restricted to fifty yards, *Challenger* was suddenly faced with the prospect of collision with an iceberg. As Matkin put it, 'Nearly everyone was on deck, it was snowing and blowing hard all the time; one officer was yelling out one order, and another something else. The engines were steaming full speed astern, and by hoisting the topsail, the ship shot past in safety.' After such experiences, the wonder was not the number of crew members lost through desertion ashore at Australasian and other ports where 'the blessings of the land' stood in especially sharp contrast to 'the raging of the sea', but the number who remained aboard to continue the trip. That they did so is a tribute to the stout-hearted example set by Captain Nares.

It is clear from contemporary accounts that Nares's leadership and navigational skills were of the highest order. He was the Ernest Shackleton of his era, and it came as a great shock to the ship's company when, after arriving in Hong Kong, Captain Nares was ordered by the Admiralty to return to England with Lieutenant Pelham Aldrich to take command of a forthcoming Arctic expedition. As it was, his replacement, Captain Frank Thurle Thomson, then commanding the frigate HMS *Modesty* on the China station, proved not to be the tyrant of rumours overheard by Joe Matkin but a highly professional officer whose skills all aboard grew to respect. The immensity of the

Captain George Nares R.N. first to command the *Challenger* expedition

HMS *Challenger* at St Paul's Rocks near the Equator
in the Atlantic Ocean

Pacific lay ahead and when at last, after three and a half years and
69,000 miles, *Challenger* lay alongside in Portsmouth harbour, Wyville
Thomson and his scientific colleagues had every reason to be content.
The Royal Navy had done them proud and, over the next twenty
years, the debt would be repaid in those fifty volumes that comprised
the final *Report* of the expedition. Sadly, Wyville Thomson, worn
out by the hardest of sea service, died before even the first volume
was published. Fortunately for the world of oceanography, the great
task was completed under the direction of Wyville Thomson's
short-tempered but academically brilliant colleague, John Murray.

One of *Challenger*'s more important discoveries was the mid-
Atlantic ridge; in making it her scientists had unwittingly encountered the junction between two tectonic plates at which new crust is
created. This is the place where the bed of the Atlantic is pulled none
too gently apart at around a thumb nail's width per year. Had they

been able to see what lay far below their keel, they might even have seen the hydrothermal vents with their luxuriant fauna based on primary production created by sulphur bacteria whose metabolism offers insights into the very origins of life itself. In the same way, when Joe Matkin cast his eyes across the Torres Straits to look with wonder at the dark volcanic mountains of New Guinea, how astonished he would have been to hear that, far below their solid bulk, the earth's crust was sinking back into its molten mantle to help balance the effects of crust formation elsewhere and thus preserve the dimensions of our fragile spaceship, Earth.

The greatest achievement of the iron men crammed into the wooden hull of the *Challenger* was to lay the foundations of modern oceanography. Those foundations have since become a mighty edifice, and it is in the light – sometimes strong and, unsurprisingly, sometimes rather more fitful – cast by this vital science that we can now study the marine lives of salmon.

To Sea with the Smolts

FOR my parents' generation, mention of the little north-east
Scottish town of Invergordon meant only one thing, the
mutiny over two days of September 1931 of the Atlantic Fleet,
then operating from their temporary base in the Cromarty Firth.
At that time, Mr Ramsay Macdonald's National Government was
facing the grim reality of the Great Depression, so severe a downturn
in the nation's economy that large cuts in public expenditure were
inevitable. The Royal Navy was not exempt but, rather than reducing
the size of a fleet already dangerously weakened by the wholesale
scrappings of capital ships and cruisers that followed the First World
War, the Admiralty elected to cut the pay of some ratings by more
than the 10 per cent suffered by the rest of the public sector. The
result was that over a thousand sailors refused to take their ships
to sea and, for the first time since Trafalgar, the sure shield of the
Empire was lowered. At length order was clumsily restored, the Rear
Admiral in temporary command was unjustly censured for his initial
leniency and, not long afterwards, the Atlantic Fleet was renamed
the Home Fleet in the nonsensical political hope that the change of
name would erase public recollection of an episode more shameful
to the bureaucrats who provoked it than the usually jolly 'jacks' who
took part in it.

Over six decades later, there are only a few coastal rescue launches
moored alongside a mole, minnows compared with the mighty
battleships and battlecruisers of 1931. Out in the firth, a small rust-

streaked trawler tows its net toward the Sutors, the low headlands
that have guarded the entrance to the firth since the glacier's final
retreat ten millennia ago. Unusually, the net is rigged to fish with its
headline breaking the surface and the trawler's ancient diesel engine
strains to tow it through the calm sea at over three knots. Up in
the wheelhouse the skipper, Alex Simpson, devout scion of a famous
north-east fishing dynasty, listens to the familiar words of *The Daily
Service*, secure in the knowledge that the net will not be hauled until
after the last notes of the BBC Singers have died away on the light
airs of mid-May. Salmon smolts migrating just below the surface
toward the broad expanse of the Moray – always pronounced 'Murry'
by those who know their Scotland and spelt as such on old charts –
Firth are the quarry, and none can be sure that our fishing matches the
short and weather-sensitive period over which the annual movement
takes place.

A glance at the wheelhouse clock reveals that the net has been
down for half an hour and it is time to haul. Apart from a solitary
cormorant skimming down the tideway and a noisy group of Arctic
terns plunging for sandeels beyond the Sutors, there are few birds
on the wing and none is taking an interest in the folds of net as the
winch's whirring barrel draws the warps steadily toward the stern.
For the kindly boatswain, long used to the curious ways of scientists,
yet another blank haul will be nothing new. At least there will be no
torn meshes to repair. How long it all seems to take before the doors
– the port and starboard otter boards that hold the wings of the net
apart – are secured astern, the extra floats are unshackled from the
wing ends and the net begins to wind itself onto the straining drum.
Here and there a pale jellyfish, softly tinted with electric blue, slides
from the meshes and falls onto the gleaming deck. Within weeks
there will be so many that fishing in the surface will be impossible. At
least the stinging cells this species uses to paralyse its planktonic prey
are too weak to penetrate the deckies' work-hardened hands. To the
deckies, the jellyfish are a messy nuisance, but to the scientist they are

The birds are, cormorant (left) and shag (right); note the white patch under the wing of the former, hence the old Scots name for the cormorant, 'letter o' marque'

a beautiful diversion from the weightier concern that his hard-won ship time will prove fruitless and he will have nothing tangible to show for the money spent on pay for the skipper and crew and diesel fuel for the ever-thirsty ship.

The buoyant cod end, the closed bag at the tail of the net that retains the catch until it is released by undoing a special knot, has been dancing lightly in the wake. Now it swings in over the gently lifting stern and, before the boatswain can untie the knot, two bright silver fish fall through the taut meshes onto the deck. A bare four inches long, they are too small to be salmon smolts but too large to be the tiny lantern fish called pearlsides, *Maurolicus muelleri* (Gmelin), which have adipose dorsal fins and look for all the world like fairy salmon when, after an onshore gale, their glittering bodies are freshly washed up along the beaches of north-east Scotland. The little fish

Three-spined sticklebacks, *Gasterosteus aculeatus* (L.)

are three-spined sticklebacks, *Gasterosteus aculeatus* L., seafaring members of exactly the same species as the tiddlers that so often breathed their last in the jam jars to which our childhood fishing confined them. By tiddler standards, the silver sticklebacks are enormous in the way that most sea trout are larger than most brown trout of the same age. The rich feeding they have enjoyed in the Moray Firth has not only paid for generous growth but also provided an unlimited supply of calcium to strengthen their formidable spines and the plate armour of scutes that partially cover their flashy sides. Now, in the late spring, they are making their way up the Cromarty Firth to seek fresh or brackish water in which to spawn. As with the sea horses and pipefishes to which they are related, the maternal duties of the female three-spined stickleback are confined to egg-laying. The hard-working male makes the nest, fertilizes the eggs and guards them and the fry until they are ready, with St Paul, 'to put away childish things' and make their own way in a world full of predators.

A firm tug at the resisting cod-end knot and more sticklebacks lie flapping stiffly on the deck. They are not alone; a softer-clad company of slightly larger fishes tumbles about our feet, their loose scales glistening like specks of chromium on the toes of our sea boots. We have caught our first smolts and, whatever else happens on the research cruise, we no longer have anything to fear from the bean counters ashore. Gently, we gather the fragile smolts and transfer them to a large tank on deck. There they swim as if still migrating, darker patches on their blue backs showing where contact with the meshes of the net and probably also with the jellyfish caught with them, has damaged their delicate skins. Now their skins are too leaky for them to survive long in the sea, so one by one we take them into the ship's small laboratory to learn as much as we can from their sacrifice. Fortunately, the ship is steady enough in the sheltered waters of the firth for us to examine their scales under the microscope. Like the rings of a tree, these little discs of bone are laid down in discrete rings at intervals that reflect the rate at which the fish is growing at the time. They grow slower in winter than in summer and so it is possible to tell the age of the fish from their scales. Interestingly, although we had caught them at the same time, not all of the smolts are of the same age. Some have left the river after two years of freshwater growth and most of the rest after three. There are even a few, mostly slightly smaller ones, that have achieved smolthood after spending just a single year in fresh water. Spreading the numbers of smolts that go to sea each year over more than one brood year helps to safeguard salmon populations against the effects of short-term problems in fresh water, especially ones created by droughts and floods. It is one of the many reasons for the great success of a species that is otherwise rather demanding in what it requires from its environment.

Nearly all of the smolts had been feeding hard on insects not long before we caught them in the late afternoon. Some of the more digested

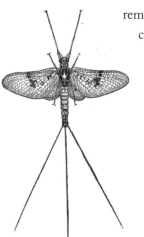

Mayfly imago

remains were of the larvae of may- and stoneflies, creatures that had grown up in the river that the smolts had shortly left behind. The fresher remains were mainly of terrestrial insects, including greenbottle flies and small beetles blown into the firth by the strengthening wind and the small shrimp-like crustaceans that marine biologists call amphipods. Clearly, the increasing saltiness of the water had not spoiled any fishy appetites. By the time we caught up again with the smolts – correctly termed 'post-smolts' now they were fully at sea – in the choppier sea beyond the Sutors, they were feeding entirely on surface-living marine animals, including post-larval sandeels that were so numerous that their gleaming bodies cascaded from the meshes to carpet the working deck with steely slivers every time the net came aboard. The post-smolts were not evenly spread across the Moray Firth but occurred in ones and twos and as small shoals. We often caught them with the baby sandeels but not when their concentrations were really dense; perhaps the post-smolts preferred not to lose contact with one another among the wriggling swarms. Every so often we would run into the sticklebacks again, a homely reminder of the days I spent fishing for them with my brothers and cousin in the River Chess. In the evening, as the gloaming drives the post-smolts to sound depths that our surface net cannot sample, we set a bridge watch and wearily turn in.

I wake to the sound of the main engine, jump into my sea overalls and, firmly grasping both rails, haul myself up the companionway that leads to the bridge. The skipper and I have every reason to be content. We set out to catch salmon smolts and that we have done on every day of the trip; we have also taken the occasional sea trout, though none has been quite large enough for the galley. We have a full day left before returning to Invergordon to refuel, and we both want to make the most of it. It is time to take stock. We have no means of tracking the smolts, which, at least in daylight, swim too close to the surface to be distinguishable by the ship's sonar from the acoustic 'clutter' created by the action of the waves. Even when they sound with the gloaming, we have no way of telling the smolts from the far more numerous other fishes spread throughout the water column. During the cruise, my colleague George Slessor, a most experienced fishery hydrographer, is keeping a detailed record of the conductivity (from which he calculates the salinity), temperature and depth of the water beneath our keel. Perhaps, we hazard, the smolts are aware of these oceanographic features and might therefore distribute themselves accordingly. But no, they cross salinity, temperature and depth contours with apparent indifference in what appears to be a headlong progress out of the firth. So it is that we decide to spend the remaining twenty-four hours pursuing the statistician's favourite pastime, repeating our observations of the days before.

After a brief return to the inner firth to refuel and to enjoy some hearty Sabbath hymn-singing courtesy of the Church of Scotland congregation of Invergordon parish, we are optimistic next morning. The shipping forecast is encouraging and a call to colleagues ashore who are monitoring the smolts leaving the North Esk in eastern Scotland confirms that large numbers are still leaving the river. Our plan is to steam close inshore toward the entrance of the North Esk, fishing as we go, in the process encountering smolts leaving the estuaries of other great east coast rivers. As we shoot the trawl in the approaches to Spey Bay, a lively anticipation shines even in the dark eyes of the normally taciturn skipper. The Spey is one of Europe's premier salmon rivers and, recalling from Sunday's sermon how 'The Assyrian came down like the wolf on the fold', we think that we have Spey's massed smolts at our mercy. We will fill the net with their delicate bodies and lay bare their lives in a feast of

Part of a catch of ninety-nine post-smolt salmon taken in the
Faeroe/Shetland Channel by FRS *Scotia*

scientific results. It takes half an hour to cross the bay, just the right length of time for our experimental haul. All is prepared in the ship's laboratory – packets for the scales, a compensated balance for weighing the fish and the contents of their stomachs, and even little tubes for the muscle samples from which our geneticist colleagues would extract DNA. We are as ready as we can ever be.

The limp cod end flops onto the deck planking, and we are reminded yet again that wild fishes have but little interest in the scientific literature. The loosening of the cod-end knot releases a single sea trout post-smolt, a larger and rather duller-hued fish than the young salmon we were after, but a welcome saviour nevertheless from the humiliation of 500 horse power's worth of blank haul. By the end of the afternoon, we have a handful of young sea trout and just two post-smolt salmon. Perhaps the Spey run was earlier than that of the North Esk, which does not enjoy the benefit of the warm Atlantic water that licks round the north of Scotland to cheer the coast of Moray. We cannot afford to spend any longer finding out but set course for the North Esk where we know that there must be fish – except that when we get there, there aren't any. The story of Spey Bay is exactly repeated and it is a glum company that sits down to dinner that evening.

At least the weather is holding. That nasty echo of the equinoctial gales that the Scotch – an old-fashioned adjective still widely used in the north-east of Scotland – fishermen of the Costa Granite call 'The gab o' May' shows no sign of wanting to add to our misery. We decide to take advantage of the good weather and, leaving the lee of the land far astern, seek the migrating salmon in the blue-grey wastes of the North Sea. We are not disappointed. At first we mainly catch sea trout, but every so often in the open sea we take little groups of post-smolt salmon. At length, we turn north and steam until we are fifty miles off Cruden Bay, once the holiday haunt of Bram Stoker, author of *Dracula*. We intercept more young salmon offshore, but the closer we get to the coast, the more the sea

The author in Arctic waters, Spitzbergen in the background

trout predominate. We even catch a single bass, at one time a rarity this far north but increasingly frequent around all of our mainland coasts in the warmer climate of today.

Our last haul before we turn for 'The Broch' (Fraserburgh) and home is within sight of the golf links at Cruden Bay. A mighty thrashing announces the arrival of the cod end on deck. At first we take the lively captive to be a returning adult salmon, but a closer look reveals what looks like a massive sea trout that might just possibly be a hybrid. Back ashore, geneticist colleagues confirm the purity of its sea trout breeding. Weeks later it takes pride of place at the laboratory's summer barbecue, where its celebrity proves short-lived. Like a lot of large sea trout that have spawned several times, its stringy flesh, which is of the dull beige hue popular with the manufacturers of hearing aids and medical underwear, is signally lacking in flavour. What a pity we did not return it the minute it came aboard.

Like a lot of research cruises, which always seem to end just when things are getting interesting, this one asked as many questions as it answered. In retrospect there is little doubt that we were probably too late to catch the peak of the salmon smolt run, but the conclusion that their feeding is not greatly interrupted as they cross the boundary between fresh and salt water now seems to be correct. There also seems little doubt that migrating salmon smolts head rapidly for the open sea, at least on this side of the Atlantic, but those of the sea trout are content with a less adventurous life in coastal waters. Why is it that these two closely related fishes use the resources of the sea so differently, despite the fact that both species go to sea for the same reason, namely to grow more rapidly than they could have done had they remained in fresh water? The answer seems to be that they are exploiting two quite separate marine habitats where the feeding opportunities are especially rich. We have 'asked the fish', and now it is time to ask the sea itself.

FAT IS A FISHY ISSUE

IT was, I think, Dr Samuel Johnson who described going to sea as like going to prison but with a chance of being drowned. That anyone should want to go to sea for fun certainly astonishes me and most of my fishermen shipmates whenever we pass a yacht on a poor day in northern waters. Salmon and sea trout are no different. Like fishermen, they brave the dangers of the sea to make a living, a much better living moreover than they could have made had they stayed behind in the place that gave them birth. As a result they are able to grow larger quicker so that, when it is time for them to breed, they are able to lay and fertilize more and larger eggs and so out-compete fish that had never left the security of the river. However, as the men of the *Challenger* expedition were able to demonstrate, the oceans' rich resources are not uniformly distributed. Over time, the ruthless forces of natural selection favour those fishes best able to target places where their food is most concentrated. Salmon make for subarctic waters enriched by nutrients from the deep ocean, and sea trout favour inshore waters enriched by nutrients from land run-off because, on the whole, there is more to eat in such areas than in the places in between.

Finding out why the seas are so patchy in the distribution of their bounty is the business of the biological oceanographer. Just as on land, there are two main sorts of living thing. The most important are the so-called primary producers, organisms like Cyanobacteria and diatoms able to support their growth and reproduction from the chemical raw materials around them, and a secondary category of less self-sufficient creatures like zooplankton – literally, animals that drift – and fishes that fatten either on the primary producers themselves or on the living and dead bodies of other members of marine life's second division. Synthesizing living material from nutrient chemicals requires energy, and one of the best ways of releasing it is through the process that chemists call 'oxidation' or, more generally, the taking away of the subatomic negative particles called electrons.

When life on earth began nearly four billion years ago, the first living things would have obtained their energy from chemical sources alone. Plenty of modern micro-organisms – for example, the sulphur bacteria that make salt marsh mud reek of rotten eggs – still do. However, only so much energy can be released in this way. Freedom from this restriction came when some early bacteria evolved the ability to photosynthesize energy-rich sugars from carbon dioxide and water. The ultimate source of the energy was light, itself a product of the nuclear fusion reactions that take place in stars and, overwhelmingly in our case, our nearest one, the Sun. From that moment on, the scene was set for the colossal expansion in the numbers and diversity of living things that we know as biological evolution and that one day would cause salmon to seek their fortunes at sea.

Of course, tapping into a virtually unlimited source of energy did not mean that henceforth biological production would be unlimited. New limits were set by the availability of the chemical building blocks that organisms need to construct their bodies. The nitrates and phosphates familiar to the modern farmer as chemical fertilizers are important, but so too are carbon to build carbohydrates, fats and proteins, silicon to make the glassy skeletons of diatoms, calcium to do the same job for coccolithophores (the main constituent of chalk) and a host of other nutrients in smaller amounts, including the magnesium needed to make the chlorophyll pigment on which photosynthesis depends. Further limits are set by the amount and availability of sunlight. Seasons and latitude set the former and

Fresh-run hen sea trout

turbidity sets the latter, since suspended particles stirred up from the sea bed or 'blooms' of algae exploiting the nutrients reduce the depth to which sunlight will penetrate. Thus the primary productivity of the sea at any one time and place is set by the complex interplay of nutrient and sunlight limits.

As sea trout have discovered, one large source of nutrients is the run-off from land. Rivers leach minerals from the soils and rocks in their catchments and this is one of the reasons why inshore waters can be so productive. Another is that winter storms stir up nutrients present in the sediments of the continental shelf and thereby make them available during the sunlit months of the subsequent growing season. Neither source of nutrients is available over large areas of the deep ocean but there are important exceptions, areas where deep water, rich in nutrients released by the decomposition of dead plant and animal material that in the past has rained down from above, is brought to the surface where sunlight can work its photosynthetic magic. One such place is in the North Atlantic where, at one of the key turning points of the global thermohaline circulation of ocean currents, the polar temperature regime causes surface water to sink. It is replaced by more saline, deep water rich in life's raw materials. Add to that bounty the long periods of summer daylight and the result is biological production on so huge a scale that even great whales migrate thousands of miles north to exploit it. They are not alone, and among the voracious creatures that share their dining table are Atlantic salmon from both European and North American rivers.

If all that grew in the surface waters of the sea were algae and bacteria – the so-called primary producers – salmon would starve there. Both they and sea trout are as carnivorous as lions and tigers are on land. Salmon and sea trout eat other animals living in mid-water. Some are members of the zooplankton, small, drifting creatures dependent for their living on the primary producers they feed upon. Some of the larger members of the zooplankton are themselves carnivores and feeding upon them are larger organisms like small fishes. The wide variety of small creatures that comprise the zooplankton are either very short-lived or, in the case of the larvae of larger animals like lobsters, crabs and fishes, spend but a small time there. So, to make a consistently good living from such a constantly changing menu, it pays not to be an over-fussy eater. Salmon cope well at sea because of their opportunistic feeding behaviour, which protects them from being too dependent on any one species of prey. Wind-blown insects, crustaceans, molluscs, worms and small fishes all feature in their diet at different times, but it would be a mistake to conclude that salmon have no dietary preferences. They seem to be particularly keen on amphipods (shrimp-like creatures also common in fresh waters), krill and small fishes. As the salmon grow, so fish play an increasingly large part in their diet, but even the largest individuals still eat other organisms, including wind-blown insects when within range of the coast. The fat-soluble carotenoid (literally, carrot-like) pigment in the amphipods, krill and, to a lesser extent, the fish is the source of the pink colour in the flesh of salmon. None of these animals can manufacture

North Atlantic krill, an important food item for salmon

the orange pigment, which is called astaxanthin. It is produced by microscopic algae and accumulates in the fat of herbivores and the animals that eat them.

The fat in their prey is very important to growing salmon and, when it is oxidized during respiration, it plays a large part in meeting their high energy requirements. Interestingly, the oxidation of fat also produces water as a by-product. It is because of the water made available by the oxidation of the fat stored in their humps that camels are able to go for long periods without drinking. Salmon in the sea face similar risks of dessication to desert mammals, and it is just possible that water derived from the oxidation of fat also helps in a small way to compensate for an environment where, in the words of the Ancient Mariner, there is 'Water, water everywhere, nor any drop to drink'. In salmon, any surplus fat is stored in muscle and connective tissues and under the skin. Later in life, this energy reserve sustains the salmon after it returns to the river, where it may remain for many months without feeding before it has to meet yet another energy-intensive cost, that of breeding. As the accumulated fat is used up, so the astaxanthin is released from the fat reserves. Some of it is lost at that time but much is redeposited elsewhere. In the hen fish, most of it ends up in the eggs. In the cock fish, it goes into the skin as the basis for the characteristic breeding livery of male salmon that seems to play a part in intimidating rivals.

Astaxanthin is, of course, added by fish farmers to the diet of intensively reared salmon but, contrary to popular opinion, it is the identical compound to the carotenoid pigment naturally present in the prey of wild salmon. In neither instance does the pigment contribute to the taste or texture of the flesh, which depend on other components of the diet, on the feeding regime and on the opportunities the fish have for active swimming. It is possible to reduce the high cost of feeding salmon by increasing the fat content of the diet at the expense of protein. Fish held at high densities and fed ad lib in this way tend to have soft, greasy flesh and bear little culinary relation to their wild-caught counterparts. The flavour tends toward the uriniferous with lower notes of stale fish food. Such fare can sometimes be difficult to dispose of. Indeed, my Jack Russell terrier Dinah, an otherwise enthusiastic little scavenger of generally catholic tastes, would refuse such an unappealing offering after a single desultory sniff. Fortunately, not all cultivated Atlantic salmon are reared with so little regard for the welfare of the fish and the subsequent eating qualities of their flesh. Indeed, a number of distinguished culinary experts now regard the peat-smoked salmon and sea trout sympathetically reared at low densities on a carefully balanced diet in the island of North Uist as the finest available anywhere in the world.

Both salmon and sea trout grow fast by exploiting the resources of the sea in regions where they are richest. In European waters, the salmon grow faster because their distant dining table in subarctic seas is more lavishly furnished than the sea trout's near-shore feeding ground. Certain North American populations of Atlantic salmon provide an interesting exception. Geneticists tell us that the species is comparatively new to that continent – invading from Europe some six hundred thousand years ago during a warm period when the Arctic ice cap virtually disappeared. It was too great a journey for the sea trout so the Atlantic salmon had the east coast of North America to itself. Over time, some of the salmon populations evolved to exploit the inshore niche occupied in Europe by the sea trout. What links the feeding behaviour of all of these enterprising animals is the capacity to make the most of what is available through not being too fussily conservative. As Man's populations expand, it is a lesson he can no longer afford to ignore.

OF BUGS AND BROOD STOCK

EVEN after Sir Robert Peel repealed the Corn Laws in 1846, so permitting the importation of cheaper wheat from abroad, poor nutrition was the lot of all too many working families during the heyday of the Industrial Revolution. Protein was in especially short supply to those unfortunates who could not afford meat and lived too far from the coast to have access to cheap fish like herring. The result was that children failed to thrive, a high proportion died and, of those who survived to adulthood, many were prevented by dietary deficiencies from reaching their full growth potential. Surrounded by servants, travelling by carriage and fed on the best their estates had to offer, it was all too easy for members of the ruling class to ignore the plight of the seriously poor. Some even attributed the suffering of the working classes to their own fecklessness, especially an unwillingness to try unfamiliar foods, in other words to embrace the opportunistic feeding behaviour that so effectively sustains the salmon in both fresh water and salt.

One of the most eccentric of these would-be dietary innovators was a certain Vincent Holt, who, in introducing his readers to the booklet he brought out in 1885, *Why not eat insects?*, tells us, 'In entering upon this work I am fully conscious of the difficulty of battling against a long-existing and deep-rooted public prejudice. I only ask of my readers a fair hearing, an impartial consideration of my arguments, and an unbiased judgement. If these be granted, I feel sure that many will be persuaded to make practical proof of the expediency of using insects as food.'

With the working classes, Holt felt that he would be pushing at a half-open door. 'After all, there is not such a very strong prejudice among the poorer classes against the swallowing of insects, as is shown by the survival in some districts of such old-fashioned medicines as wood-lice pills, and snails and slugs as a cure for consumption. I myself also knew a labourer, some years ago, in the west of England, who was regularly in the habit of picking up and eating any small white slugs which he happened to see, as tidbits, just as he would have picked wild strawberries.'

A large land snail of the type recommended for working people's tables by the author of *Why not eat insects?*

Holt's idea of what constituted an insect was decidedly plastic because, as we have seen, it included spiders, woodlice, slugs and snails as well as true insects and their larvae. However, if his zoological taste was catholic, he insisted that only vegetarian 'insects' should find their way onto the plates of the deserving poor. Just as he would not expect them to eat the flesh of such carnivores as 'cat, dog or fox', so he did not 'ask my readers to consider for a moment the propriety or advisability of tasting such unclean-feeding insects as the common fly, the carrion beetle or indeed the churchyard beetle'. Clearly, it had not struck him that virtually all of the fish of commerce are carnivores.

Warming to his theme, Holt appealed to his comfortably placed fellows to lead the culinary way.

How the poor live! Badly, I know; but they neglect wholesome foods from a foolish prejudice which it should be the task of their betters, by their example, to overcome. One of the constant questions of the day is, how can the farmer most successfully battle with the insect devourers of his crops? I suggest that these insect devourers should be collected by the poor as food. Why not? I do not mean to pretend that the poor could live upon insects; but I do say that they might thus pleasantly and wholesomely vary their present diet while, at the same time, conferring a great benefit upon the agricultural world. Not only would their children then be rewarded by the farmers for hand-picking the destructive insects, but they would be doubly rewarded by partaking of toothsome and nourishing insect dishes at home.

Disappointingly for Holt, his kindly meant suggestions were not taken up – even, I suspect, by the author himself, whose insectivorous tastes were almost certainly rather less well developed than he would have had his readers believe. Frank Buckland was another dietary innovator, but although, as we have seen, he extolled the wholesome qualities of toasted field mice and young rats and, as a child, had tasted a showman's crocodile – the flavour resembled that of 'broiled lamp wick' – that had sickened and died, his approach to broadening the national diet was decidedly more realistic. He would never recommend to others species he had not sampled himself. He also had extensive contacts with gamekeepers, fishermen, bird catchers and animal keepers, men who knew the difference between unfamiliar but wholesome fare and the fantasies of drawing-room eccentrics.

Buckland was, for example, rightly cautious about the likely public acceptability of equine flesh, having heard of the poor reception accorded to the dripping that ran from the young donkey fattened for the table on oil cake by the Fellows of Sidney Sussex College, Cambridge. In those days, dripping was one of the perquisites of the college servants, but they would have nothing to do with it that week. Buckland's faith in the likely market for horse meat was tested more directly when he attended a dinner at the Langham Hotel in London at which 168 people sat down to enjoy a dinner in which every one of the dishes, from the soup to the jelly, was of equine origin. Buckland reported that all of the dishes had an unusual flavour. 'It reminds one of the peculiar odour which pervades the air in the neighbourhood of a horse which has been "hard galloped".' He continued:

> In the middle of dinner I stood up to watch the countenances of the people eating, and I devoutly wished I had had the talent of a Hogarth to be able to record the various expressions. Instead of 'men's beards wagging', there seemed to be a dubious and inquisitive cast spread over the features of most who were present. Many, indeed, reminded me of the attitude of a person about to take a pill and draught; not a rush at the food, but a 'one, two, three!' expression about them, coupled not unfrequently by calling in the aid of the olfactory powers, reminding one of the short and doubtful sniff that a domestic puss not over-hungry takes of a bit of bread and butter. The bolder experimenters gulped down the meat, and instantly followed it with a draught of champagne.'

What little remained of the initial enthusiasm of the diners was finally destroyed by a photograph, circulated at the end of the evening, of the broken-down old cab horse that had contributed its stringy carcass to the meal. Small wonder, then, that Buckland concluded, 'In my humble opinion, hippophagy has not the slightest chance of success in this country.' As he said at the time, 'Among the better classes, the flesh of the horse will never become popular; for in the first place, the cooks will not cook it (unless they are placed under martial law); in the second place, the ladies will object to it; and thirdly, the master of the house will find it vastly inferior to beef and mutton.'

Rather than promote species against the consumption of which there was already a strong national prejudice, Buckland believed that the culinary future lay with the introduction of the truly novel. In this ambition he was not alone. His father, William Buckland, had been one of the three founder members of the Zoological Society of London. Nowadays its activities are largely scientific and educative.

A rather stringy cab horse resting from its duties

So it was back in 1825, but in those days there was a stronger emphasis on Man the lord of creation than on Man, just another species of great ape. Thus, in an early prospectus of the Society, we read, 'When it is considered how few amongst the immense variety of animated beings have been hitherto applied to the uses of Man, and that most of those which have been domesticated or subdued belong to the early periods of society, and to the efforts of savage or uncultured nations, it is impossible not to hope for many new, brilliant and useful results in the same field by the application of the wealth, ingenuity, and varied resources of a civilized people.'

As the century progressed, so did a strong interest in 'acclimatization', defined by Frank Buckland in 1859 as 'a term which may be said to comprehend the art of discovering animals, beasts, birds, fishes, insects, plants and other natural products, and utilizing them in places where they were unknown before'. Buckland was interested in all of the biological categories listed in his definition, but his greatest personal contribution was to be among the fishes and especially those of the salmon family. One of the first fruits of the acclimatization movement was the foundation in Paris of the Société Impériale d'Acclimatation in 1854. The success of the new organization was immediate and by 1860 it had recruited two thousand members including 35 royal personages. Among the latter were the French Emperor, the Pope, the Emperor of Brazil and the King of Siam. Such an important and successful initiative by the British Empire's principal colonial rival could no longer be ignored. The intention to form an Acclimatization Society was advertised in *The Field* in the same year. Distinguished patrons were soon rounded up; the Marquess of Breadalbane was elected President and Frank Buckland, Secretary, with the addition, soon after, of the linguist James Lowe, one of Buckland's fellow contributors to *The Field*.

This is not the place for a detailed treatment of the Society's many projects, involving as they did such exotic creatures as the eland, the guinea pig-like capybara (the largest rodent in the world) and the great Australian kangaroo. As a certain Dr J. E. Gray said at the British Association's annual meeting in 1864, 'Some of the schemes of the would-be acclimatizers are incapable of being carried out, and would never have been suggested if their promoters had been better acquainted with the habits and manners of the animals on which experiments are proposed to be made.' Such an accusation could fairly have been made against Frank Buckland when, during an address to the Society of Arts in 1860, he advocated stocking the parks of the seriously rich with obscure alien ungulates of all sorts, of which the larger species included reindeer, wapiti (a giant North American relative of the red deer), moose and yaks. He was less well disposed toward kangaroos, not because he saw anything wrong with their 'venison' but because 'it would not suit the rural scene'.

Enthusiastic as Buckland was about the domestication of unfamiliar mammals, he was much surer of his biological facts when it came to the culture of salmon and trout and it was to this activity that he was to make his most lasting contribution. At the time it was seen both as a means to augment the domestic stocks and as a way of extending the geographical range and abundance of two of the most highly prized fishery resources in the home country to some of the farthest corners of the British Empire. Nowadays, however, releasing even the most valuable of alien species into the wild tends to be frowned upon by conservationists. The principal modern applications of the techniques pioneered by Buckland and his contemporaries are to intensive cage and pond culture and, at a time of reduced natural

Inside an early hatchery in the days when salmon culture was a pastime for smartly turned out gentlemen!

survival, they may also serve as a way of increasing the numbers of fish available to the angler.

As a classical scholar and Wykehamist, Buckland would have been well aware of the ancient use of stew ponds to rear carp and possibly of the obsessive interest of Roman epicures in large red mullet. He would probably also have known that the artificial spawning of trout had been practised by the French monk Dom Pinchon as early as the fourteenth century. Buckland himself was brought up to date with the latest developments in salmon culture following the visit of his father William to Drumlanrig Castle, seat of the Dukes of Buccleuch, in 1844 in company with his friend Louis Agassiz, doyen of French ichthyologists. His Grace's gifted head keeper, John Shaw, had made a detailed study of the early life of salmon and had reported the results in a series of scientific papers to the Royal Society of Edinburgh. One of the breakthroughs in the successful artificial spawning of salmon was the discovery that fertilization was most successful when the eggs were expressed into a clean, dry bowl and the milt stirred in among them without the previous addition of the water always present when salmon spawn naturally, burying their eggs in redds in river gravels. The explanation for this counter-intuitive observation, which may or may not have been known to Buckland, who often obtained his eggs ready-fertilized by excavating naturally produced redds, lay in the curious fact that fresh water is quickly lethal to the spermatozoa of salmon, which survive best in brackish water. It was not until halfway through the twentieth century that Dr Jack Jones of the University of Liverpool showed how the joint participation of large cock sea-run salmon and small male parr that had achieved sexual maturity in fresh water ensured that egg and sperm were brought into intimate contact before the sperm became waterlogged and died.

However Buckland obtained his fertilized eggs, 'for stocking salmon and trout in places where they did not exist before', he was the first scientist to state in writing that 'the judicious cultivation and opening up of natural spawning grounds will, in my opinion, be

Frank Buckland holding a tile used to collect oyster spat

always preferable to hatching the Salmonidae by artificial means'. Only in sometimes recommending stocking, 'to improve the breed of fish', did the views of this pioneering Victorian differ from those of the modern fishery scientist. Because, as Buckland himself believed long before it was proved experimentally, the Atlantic salmon is a homing species; it tends to form locally distinct populations, even within single river systems. Some of the differences created in this way are adaptive, that is to say they improve the survival and reproductive opportunities of the individual fish that carry them. The size and number of their eggs, the rate at which the young salmon develop and the timing of their migrations as both smolts and adults are examples of characteristics sensitive to local natural selection. Recent research in Spain and Ireland has demonstrated beyond doubt that locally adapted fish out-perform those introduced from elsewhere. In the Spanish experience, local fish out-performed those introduced from elsewhere in terms of both freshwater and marine survival. The same applied in Ireland apart from individuals from one of the alien sources. Rather disconcertingly, these fish easily out-competed the depleted local population in fresh water only to fail hopelessly at sea. The effect of introducing them was therefore to reduce even further the numbers of adult salmon returning to the river.

Buckland's own efforts to secure brood stock were truly heroic, as the following diary entry indicates:

My water-dress put on, the nets and cans, &c. packed, we started in a carriage, with a pair of fast horses, for a brook, which we calculated would not be running so much water as the main river. Our theory was right; but still there was too much water to render netting anything but very hard work. A pair of salmon had been marked down the day before; so taking out the pole of the pole-net (reader, imagine an old-fashioned, bag-shaped night-cap, with a stick fastened on each side of it, and you have a pole-net), I attempted to cross the stream, at a spot below where the salmon were supposed to be, and, with great difficulty (on account of the huge stones and tremendous stream, about an inch over the waist), I got across, and then, sticking the iron-shod end of the pole into the bottom of the river, I got to the lower side of the bank. My friends and the keepers then went about eighty yards up the stream, and began to beat it down, and to throw a shower of stones into the rapid, foam-coloured brook. Almost instantly I perceived coming down stream, with the rapidity of a partridge's flight as he tops a hedge, two huge bodies and two black fins at the surface of the water. 'Look out,' cried my friend above, 'they are coming down.' It was with difficulty that I could hold the empty net against the stream, but when the two great fish came rushing on, all steam up, with a double-barrelled bang, bang, against the net, the whole apparatus was nearly jerked out of my hands. The instant the fish were in the net, I tried to scramble across the brook to the other side, with the pole in my hand, so as to enclose the fish and catch them. A huge stone was in the way, and, being over-anxious, I stumbled, and the stream as nearly as possible had me down; then, to add to my difficulties, just at that moment something ran between my legs, and nearly upset me altogether. 'Oh dear! Oh dear!' I cried, 'one of the salmon is gone, but we have got the other, for I can feel it jerking in the net.' We hauled it quickly onshore, and my eyes were gladdened by the sight of a splendid hen salmon as full of eggs as could be, but remonstrating terribly in her way at the rude manner in which we were interfering with her domestic arrangements. I took her gently by the wrist, the thin part of a salmon's tail, and then let her down into the net I use for keeping fish alive, and went to work to catch her husband. We marked him down under a bank, but, when we came to try for him, he was gone, having crawled in, as it were, under the roots of an old pollard tree, from whence no amount of stirring with sticks or pelting with stones would make him budge an inch; so Sir Salmon beat us after all. We then made three distinct pitches with the nets higher up the stream. We caught two more salmon, but none suited for our purpose. It was getting dark, and we were all very wet and cold; so we adjourned the meeting till the next day.

As to the subsequent fate of the young fish stocked by Buckland, we can only regret with him that a reliable and at the same time reasonably harmless means of marking them so that they could be recognized later as returning adults, was not available to him. The stocking activities of which he was most proud were his contributions to the successive efforts to introduce trout and Atlantic salmon to Tasmania and New Zealand by shipping trays of ice-cooled ova. The trout succeeded but the salmon failed, almost certainly because the genetic programming that lay behind the dynamics of their long migrations in the North Atlantic were inappropriate to the South Pacific. Interestingly, sea trout runs, no doubt based on relatively short coastal migrations, are now well established in a number of systems in the southern hemisphere. It is possible therefore that, had Buckland obtained his fertilized ova from one of the sea trout-like Atlantic salmon populations that do not migrate beyond the Bay of Fundy in Canada, he might have succeeded with this species also. Sea migratory salmon now occur in the South Island of New Zealand, but they are quinnat (also known as chinook) salmon, *Oncorhyncus tshawytscha* (Walbaum), native to the Pacific coast of North America. Some chinook populations do not undertake long sea migrations but behave rather like sea trout. Almost certainly it was from brood stock taken from one of these coastal chinook populations that gave rise to New Zealand's only sea-run salmon population.

Perhaps the apogee of Buckland's fame as a fish culturist was reached in January 1863 with his setting up of a miniature salmon hatchery in the front window of the *Field*'s offices, at that time in the Strand and an address that ensured the maximum of fashionable attention from London society. That the day-to-day management of

Frank Buckland's salmon hatching exhibit at the
Islington Dog Show

the apparatus was in the hands of Buckland's old friend, the viper-
catcher Thomas White, added to the sense of theatre that often
attached itself to Buckland's enterprises. Buckland's lively monthly
accounts of the progress of the young fish gave substance to his
showmanship. Later in the same year, he received an invitation to
exhibit the apparatus at the prestigious Islington Dog Show where
both he and a young salmon he had with him in a 'bottle of physic'
were presented to the Prince of Wales (later King Edward VII), who
made a 'careful inspection'.

If his public reputation was largely based on his efforts to promote
the cause of salmon and trout culture, there is no doubt that Buckland

did even more lasting good by his efforts as Inspector of Salmon Fisheries to restore their rivers. Unlike in Scotland, where the private ownership of the fishings had done much to protect the rivers and where the worst excesses of the Industrial Revolution were largely confined to the Forth-Clyde valley, many fine salmon and sea trout rivers south of the border had suffered grievously. As we have seen, Buckland was, to his great credit, the first to recognize the different physical habitats required by the successive life stages of the fish, and also the fact that all these stages demanded well-oxygenated water free of industrial and domestic wastes. He was especially scandalized by one factory owner who asserted that sulphuric acid was 'a tonic for the fish'. Buckland replied, 'Manufacturers of all kinds of materials, from paper down to stockings, seem to think that rivers are convenient channels kindly given them by nature to carry away at little or no cost the refuse of their works.' He was particularly concerned to secure the uninterrupted passage of fish migrating both down and upstream and would emphasize that a heavily polluted estuary could be as big a barrier to migration as a mill dam or badly run fish weir. Thus he reserved his strongest censure for certain mine owners whom he accused of 'cutting off the very sources of the rivers and converting them into poisonous streams'. Despite his trenchant criticisms of the polluters, however, he was strongly of the opinion that the developing science of sewage treatment would one day allow industry and salmon to live in harmony. Long after Buckland's death in 1880, we have seen the fulfilment of this prophesy in the cleaning up of many of the rivers whose noisome state so concerned him. How pleased he would have been to hear again of salmon in the Kelvin, Clyde, Tyne and Thames and of the wonderful results that have attended the opening up of caulds or weirs on Tweed so that it is now the premier salmon river in Great Britain.

THE HYDROGRAPHER'S FISH

SALMON and sea trout may not be fussy eaters but, as Buckland could have told us, they are very choosy indeed about the quality of the fresh waters they require to sustain their reproductive and juvenile lives. They thrive only in the cleanest of rivers. They are welcomed not just for themselves but because their presence is seen as a reassuring indication that all must also be well with other less demanding species, varieties like pike, fish of the carp kind and eels, all of which can endure reductions in the levels of dissolved oxygen that would kill a salmon or sea trout in minutes. Fortunately, the surface waters of the open sea where they make most of their growth and where in remote antiquity their ancestors evolved, are rich in oxygen. Temperate and sub-polar seas also support high levels of potential food organisms and, like all marine waters, are protected by what chemists call 'buffering' from the acute increases in acidity that can spell death to the early stages of salmonid fishes in some otherwise pristine river systems. Above all, the seas are large. The limits to abundance set by the amount of food and space in fresh water do not apply in the same strict way in the relative freedom of the sea. Not only do salmon grow much faster there than in the river, but there is room for more, many more, than the numbers of smolts that are naturally produced. Therefore, it was believed by some biologists nearly forty years ago that a salmon 'ranching' industry could be developed by releasing artificially reared salmon smolts to sea and harvesting the returning adult fish in traps.

It was an attractive idea that would undoubtedly have appealed to Frank Buckland. Unlike the intensive rearing of caged fish that had to be fed on a relatively costly high-protein diet for the whole of their lives, the sea would provide for the ranched fish for nothing. Not only that, but the fish that survived to harvest would have all the eating qualities of wild salmon. Unfortunately for this worthy attempt at true fish farming, practical experience showed that, although the basic premise that the sea had untapped salmon-producing potential was undoubtedly correct, not enough fish came back every year to meet the costs of rearing the smolts. It soon became clear from tagging experiments that the levels of return from even the most carefully reared smolts rarely approached those of wild ones. Perhaps the 'boarding kennel' fish were deficient in the skills needed to catch live prey in the open sea or to avoid the predators that abound there. Direct observation was to provide clear evidence for both explanations.

When salmon grow in length, so do their scales, in the process laying down ridges called circuli at a rate broadly in proportion with the growth rate of the rest of the fish. Thus salmon carry with them a record of their past growth and also of their relative success in securing a living from the food available to them. Comparison by my shipmate Julian MacLean of a large sample of scales, taken from wild and artificially reared young salmon fished by surface trawl from the Faeroe-Shetland Channel some seven to eight weeks after they entered the sea, showed that, despite their larger average size on entry, the reared fish were growing at but a third to a half of the rate of the wild ones. If slow early growth is a potential death sentence for a small fish vulnerable to fast-swimming predators, so are deficient behavioural defences. My Icelandic colleague Tumi Tomasson noticed that, during their migration to sea, wild smolts form much tighter shoals than reared ones, an observation confirmed shortly afterwards in a large marine aquarium in Scotland. As many a convoy commodore would confirm, a tight shoal is more protective

Atlantic salmon scale; the circuli indicate that this fish had
gone to sea as a three year old smolt and returned to the river
after two years at sea

than a loose one because of its greater ability to take rapid avoiding
action through synchronized turning. It also leaves fewer stragglers.
Although we would never have the entire explanation for the lower
survival of reared smolts at sea, we now had enough of one to shake
our faith in the likelihood of achieving an economically viable
alternative to the intensive culture of salmon in cages.

Absorbing as comparisons between the marine performance of
reared and wild smolts can be, the recent revival of interest in the
lives of salmon at sea has been driven largely by concerns about a
sustained reduction in the return rates of fish of both types. It is
a concern shared by North American scientists, including those
who study the very different species of salmon in the Pacific. Could
it be that, just as in fresh water, salmon are sensitive indicators of

widespread environmental change – that they are indeed, as my shipmate, the distinguished physical oceanographer Dr Bill Turrell, christened them, 'the hydrographer's fish'?

The first place to look was not in the sea but at the fish themselves. Salmon return to the rivers of their birth to spawn, a migration that is triggered by the early stirrings of the same hormonal processes that in us transform the sunny charm of childhood into the grunting unattractiveness of adolescence. Just as in ourselves, this radical change does not take place at the same time in all individuals but over a range of ages that seems to be sensitive to the food available. 'Ten-year-olds having sex,' shrieks the tabloid front page without connecting this sensational news with the previous week's headline, 'Childhood obesity, Britain's kids the fattest in Europe'. So it is with salmon, if there is enough food to fuel both bodily growth and the laying down of rich fat reserves, sexual maturation, and therefore the stimulus to return to the river, tends to happen at a younger age than at leaner times. Such a response to better feeding opportunity tends to be favoured by natural selection because, although it shortens the total lifetime of the fish, it also shortens the period over which it is at risk of dying before it has had the opportunity to breed.

The last time the salmon was favoured in this way seems to have been in the 1960s and early seventies when, during a period of cooling in northern waters, the levels of plankton production in the subarctic mixing zones where salmon make most of their growth were unusually high. The result was that the salmon grew fast and, although most returned after only a year at sea as what the stained-glass language of the fish trade calls 'grilse', they were big for their age and were coming back in exceptionally large numbers. The lavish feeding opportunities also promoted the increased survival of those fish that had not been triggered to return until after their first or second years at sea. So, although these older salmon formed a minority of the total stock, the numbers of these especially valuable

Grilse, a salmon that has returned to the river after spending
just one winter at sea

fish entering home waters four decades ago were still much higher than those available to us today.

Paradoxically, grilse now form an even greater proportion of the total stock, but not because good feeding has promoted an early puberty. For, along with the decline in the numbers of returning fish, there has also been a sharp reduction in the proportion, always small, of the stock that survives a further period at sea to return and spawn twice or even three times. The exceptional rarity of these 'previous spawners' was further direct evidence of the increased rate of mortality that salmon of all ages are currently suffering and that has its most withering effect on the unfortunates exposed to it longest. Here, it seemed, was the main reason why the reduced numbers of returning fish consist mainly of grilse.

Although studying returning fish had told us much, we could not hope to discover what might have been the fate of those that did not return without rejoining them in their element. So far, we had satisfied ourselves that the salmon that left our eastern rivers did so rapidly, crossing salinity and temperature barriers in the North Sea with apparent abandon until they were out of range of our aged inshore research vessel, which at times seemed to us to be connected to its home port on the Costa Granite by a gigantic piece of elastic. Before we lost contact, we confirmed again that the little fish migrated in small shoals, feeding as they did so on larger items in the surface zooplankton. Krill, other crustaceans and the post-larval stages of fishes were the main targets of the young salmon and, among the fish prey, the young of the common sand eel, *Ammodytes marinus* Raitt, featured strongly.

Of course, the numbers of young salmon, constrained as they were by the rearing capacity of their parent rivers, were far too low to pose much of a threat to the sandeels. Not only that but it appeared to us that they took their victims from the edges of the shoal, perhaps because entering its dense core would cause the young salmon to lose sight of the black-edged fins of their fellows, so leading to the break-up of their own protective aggregations. Valuable as the massed shoals of sandeels clearly were to our post-smolts in fuelling their early marine

Adult sandeel

growth, the ready availability of this rich food supply appeared not to delay their rapid migration north. Just as their internal navigational computer was able to drive the young salmon across temperature and salinity boundaries, so it seemed it could also pilot them through a swimming banquet so lavish that it would surely have held up any fish that could think for itself. The fact was that the post-smolts were programmed to make their way to even greater feeding opportunities beyond the continental shelf. They were on their way to the central Norwegian Sea.

How they keep this vital tryst is still something of a mystery. It is all very well to have a brain programmed to direct its owner to ignore all distractions and swim northwards, but how does the fish know where north is? One clue was uncovered by Dr Andy Moore, now head of salmon research at the Lowestoft Laboratory. He discovered that the lateral line of salmon, which confers a directional sense of touch at a distance, also carries innervated particles of the iron compound called magnetite, a substance that as its name suggests responds strongly to magnetic forces. Not only could such a system indicate the general direction of north, but it could also provide the fish with more detailed information about local variations in the strength of the Earth's magnetic field. In other words, it would provide the salmon with access to a sort of map of what lay below it.

It was to be some years after we had completed our work in the North Sea that we saw at first hand how the salmon might use this system. By that time my Norwegian colleague Dr Jens Christian Holst had persuaded his parent laboratory in Bergen that they should let him borrow a new type of large surface trawl fitted with a television camera. It was a remarkable piece of fishing gear. For a start it was rather an odd shape in that its opening was only ten metres deep but was no less than sixty metres wide. It was designed to sample the surface waters of the sea where our previous experience had shown that migrating post-smolt salmon spend the long hours of daylight. To a commercial fisherman, the oddest feature of the net was that it could never have caught fish because it had no cod end (the name is derived from the item of knightly attire called a 'cod piece' and actually has nothing whatever to do with cod). Just ahead of where the cod end would normally be attached was a triangular aluminium frame onto which was mounted a closed-circuit television camera. It was placed at right angles to the direction in which the net was being towed through the water so that we could watch what we would otherwise have caught. Transmitting the images from the camera to the scientists on the bridge was done by radio, a method complicated by the fact that radio waves are severely attenuated by sea water. The solution was to lead an insulated cable from the camera to an aerial carried on a separate twin hull towed astern. There were two advantages to this strange method of fishing. The first was that it saved the lives of the fish; the second and more important one was that it enabled us to fish continuously without having to haul the net every hour or so to see what we had caught. As a result, we would know precisely where we were every time we encountered the salmon.

Our plan was to use the gear to follow the post-smolts from the west of the British Isles as they migrated north to the west of the Outer Hebrides and through the Faeroe-Shetland Channel on their way to the Norwegian Sea. Unfortunately, with that capriciousness that has often caused me to question the sincerity of the Almighty's commitment to the cause of biological oceanography, a 'gab o' May' depression arrived just as we left Lerwick, a poor day rapidly turned into what Scottish trawlermen call a coarse day and we were obliged to shoot the gear for the first time in the relative shelter of the Northern Minch, the strip of water that separates Lewis and Harris from the mainland of Scotland. It was probably just as well. Before we had even connected the various parts of the gear, we found that, somehow or other, sea water had found its way into the transmission cable. There, in the lee of the bleak hills of Harris, my patient hydrographic colleague, John Beaton, and that most reliable repository of electrical common sense, the staff of the ship's engine room, located the fault, cut away the shorted-out cable and presented Jens Christian and me with exactly what we needed. Perhaps the Almighty was watching over us after all, but I daresay that, if I had asked any of the deckhands, he would have dismissed this change in our fortunes as nothing more than the Devil looking after his own.

Whichever member of the celestial – or infernal - pantheon was now looking after our interests, his duties were not yet over. Before we could start fishing, we needed to be sure that all parts of the gear took up their proper positions when streamed astern over the full range of trawling speeds we intended to use. There was only one way to find out and that was to leave the security of the ship and join the towed trawl in its element. Stiff in our orange survival suits, we took our places in the RIB (rigid inflatable boat), relieved to note that the derrick had a good, if jerkily applied brake, and settled onto the long swell as it creamed past the hull now towering above our frail craft. A muffled roar from the outboard and we were clear of the ship and skimming among the kittiwakes and mollies (the fisherman's name

The Royal Norwegian research ship, *Johan Hjort*

for fulmars) gathering in the false expectation that we were really fishing. As to the trawl, the floats buoying up its headline and the little catamaran carrying the aerial, all were bobbing happily along in the surface, even when we asked the ship to tow the gear at the relatively high trawling speed of four and a half knots. Within minutes of being swung dizzily back aboard, we had seen our first two post-smolt salmon on the CCTV screen and not long after that the wind moderated and we were able to leave the Minch and make for the edge of the continental shelf to the west of the island of Barra.

The transition between the shelf and the deep ocean is known as the continental slope. It has long been of great interest to hydrographers, not least because of the northerly current that flows along it. It has been of even longer interest to post-smolt salmon, which use it to assist their passage to the Norwegian Sea. What baffled us was how fish swimming in a moving body of water at the surface, and therefore with no obvious point of reference from the sea bed far below or from coastal features many miles away, could possibly tell whether they were in the current or not. Our faith in the idea that sensitivity to temperature and salinity boundaries held the key had been shaken somewhat by our experience in the Moray Firth, where fish making their way north seemed to ignore them. However, this could simply have been a result of our crude sampling method since at that time we had only a conventional trawl and therefore did not know either when or where exactly in the North Sea the fish we caught had entered the net.

But now, fishing off the Western Isles with the CCTV-equipped trawl, we could record the position of every fish as it entered the net. We were astonished to discover how good the post-smolts were at staying within the very core of the Slope Current. Once again they seemed to ignore salinity, temperature and depth boundaries, even turning sharply across them as, in the surface, they passed over a prominent sea bed feature far below, the Wyville Thomson Ridge, to enter the Faeroe-Shetland Channel. Having tentatively eliminated

The open, CCTV-equipped cod end of a mid-water trawl used to
count surface swimming post-smolt and adult salmon

these three possibilities, we still do not know what navigational
aids these astonishing products of natural selection use during
their time in the open sea. Of the several candidates, including a
number of celestial ones, the discovery of the magnetite particles
in the lateral line suggests that an unconscious sensitivity to the
local and general features of the earth's magnetic field may well be
of special importance. Not until the young navigators had passed
between Iceland and Norway and reached the latitude of another
prominent sea bed feature known as the Vøering Plateau would they
leave the security of the Shelf Edge. From then until changes in
certain pituitary hormone levels indicated that the time had come to
return to their home rivers, they would cruise in the system of slowly
circulating currents known as 'gyres' where, in the Norwegian Sea

and in the waters of east and west Greenland, they would feed and grow at rates rivalled by few other cold-blooded creatures.

The results of our exploratory fishing also indicated that, in daylight, the young salmon spent long periods in the top few metres of the sea. Whether or not they spend all their time up there we do not yet know. By examining the output from depth-recording tags he had attached to larger salmon, Jens Christian discovered that some of them undertook feeding dives to over one hundred metres in daytime and it remains to be seen whether post-smolt salmon behave in a similar manner. The difficulty we have in catching post-smolts in subdued light suggests that they leave the upper ten metres of the sea in these conditions, although whether they do so to feed like the sub-adults in daytime or to avoid predators, we cannot tell. Such is the present extent of our knowledge. But now research teams on both sides of the Atlantic have combined their resources in a three-year international programme co-ordinated by the North Atlantic Salmon Conservation Organization based in Edinburgh and, by the end of this endeavour, we should know exactly how salmon of all sizes while away their days at sea.

By Ebrie's Bleak Banks

O N the face of it, the salmon's life in two quite different media, involving as it does migrations over thousands of miles dependent for their successful completion on both innate and learned behaviour, is so risk-ridden a way for a fish to perpetuate its kind that it is difficult to see how it could possibly have evolved. What we can be sure of, though, is that it did not do so in a single step but developed gradually over tens of millions of years, perhaps from a more recent starting point echoing the life cycle of some of the closely related sea-migratory trout populations of the present day.

I was to encounter these versatile fishes for the first time not long after making a migration of my own to the north-east of Scotland, where my new home was at the site of an eighteenth-century kirk with a remarkable history. By the sixteenth century, the institutions of western Christendom had acquired so many worldly overlays that the bright light shed by the brave souls of first-century Palestine had been reduced to a guttering flicker. For the earnest divines of Geneva, the time had come to discard the veneration of relics and holy images, literal belief in the transubstantiation of the host and, above all, the sale of indulgences. As the case for reform strengthened and spread across north-west Europe, so the conviction grew that the core values of the early fathers were best recovered, not by reference to a Papal hierarchy, but by the direct study of Holy Scripture. If the support of congregations was to be achieved, this reform required that the Old and New Testaments be made available in a language that all could understand. This is not the place to revisit the horrors that attended the Reformation and its Counter, merely to note that, however well-meant at the start, revolutions have a horrible habit of getting out of hand. In replacing the authority of Rome with that of Holy Scripture, the Protestant reformers had exchanged a single, if at times corrupt, strand of leadership with a bafflingly many-headed hydra.

By the twentieth century, the problem for the reformed clergy was to convince their flocks that the Holy Scriptures were nevertheless the revealed word of a loving God. Brought up by his tutor, John Neale Dalton, to read his Bible every day, King George V confessed that he had found 'many curious things in it'. Could it really have been a sign of Godly favour that, whereas King Saul had merely slain 'his thousands', his revered successor, David, had slain his 'tens of thousands'? Or, as my old friend and former leader of the Iona community, the Reverend Ron Ferguson, pawkily asked the readers of *Life and Work* – the magazine of the Church of Scotland – what about the story of Elisha, second only to Elijah himself among the great prophets of the Old Testament? As we read in the second book of Kings, 'And he went up from thence unto Bethel; and as he was going up by the way, there came forth little children out of the city; and mocked him, and said unto him, "Go up, thou bald head; go up, thou bald head". And he turned back, and looked on them, and cursed them in the name of the Lord. And there came forth two she-bears out of the wood, and tore forty and two children of them.' Whatever one's views on zero tolerance, on this evidence a tendency to overreact might seem to the sceptic to be one of the less attractive aspects of the Deity's complex personality. The better-educated of the Protestant divines had of course coped with such problems through the simple expedient of selective quotation. Such preachers were highly sought after, but how were their congregations to attract such paragons? In Scotland, the new Protestant churches with their elevated pulpits and austere interiors were little more than

Craigdam Kirk, the third Seceder Kirk in Scotland photographed
just prior to demolition

basic preaching boxes. The best of them have a minimalist vernacular
charm to a twenty-first-century eye, but few would regard them as
attractive buildings. Congregations competed not by offering more
elaborate churches but through the size and appointments of their
manses and glebes, the agricultural land set aside for the minister.

Back in the 1970s, I lived with my family in the remnants of the
third Seceder Kirk in Scotland, a relic of one of the more extreme
forms of eighteenth-century Presbyterianism. Despite the lack of
lordly patronage, something the Seceders eschewed, the congregation
had spared no expense in constructing a magnificent manse in their
successful efforts to secure a succession of notable preachers. The
independent life of Craigdam Kirk reached its sad end in 1958, and
the manse was sold as a private house. In the late seventies, it was
owned by a helicopter pilot, later to die tragically in a flying accident,
serving the oil platforms of the North Sea. Gerald Hardy had met his
Australian wife, Margaret, when serving as flight commander aboard

the survey ship HMS *Hydra*. Their two children, a boy called Robbie and a sweet little girl called 'Blidda' (Robbie's attempt at Elizabeth), were a little younger than our two boys, John and Neil, with whom they were great friends.

John and Neil were often to be found cooking small brown trout they had caught in the reed-fringed burn that flowed through three of our farming neighbours' fields. Both Robbie and Blidda had healthy appetites and Robbie in particular rather envied the boys' access to 'spunks', as matches are called in north-east Scotland. The smoke of a cooking fire would normally bring both of them running to join their older neighbours and it was not long before Blidda, who

Young family and friends at Craigdam; my sons, John and
Neil (standing), Robbie Hardy, "Blidda" Hardy, Fergus Perry
and self (sitting)

was closer to our boys in age, was asking about the source of the fish, and we agreed to take her on the next expedition. Rather than mount another attempt on the lives of the remaining local trout, the boys were keen to try their luck with the bigger game of the Ebrie Burn, a tributary of the River Ythan (pronounced Eyethan). The Ythan is a modest stream that tumbles through the Braes of Gight (pronounced Gecht), calf country of Lord Byron, and thence via a ribbon of remote lowland parishes to the settlements of Ellon and Newburgh. The course of the river marks the southern boundary of one of the bleakest regions of mainland Scotland, the knuckle of Aberdeenshire known as Buchan. Its rolling acres, famed for the horn and corn their fertile soil produces, are bounded, not by hedges, but by low stone dykes and their ugly modern successor, the barbed-wire fence. Trees hang patchily around some of the steadings and in larger clumps in the policies of the few remaining great estates but, over much of the peninsula, there is little to break the pitiless gales from the North Sea. Anthropologists who have studied the rounded skulls found in the Bronze Age burial sites, known as short kists from the shape of their stone coffins, tell us that similar folk are still well represented in the modern population. Such tenacity in the face of so testing a climate has left its mark on the character of these rugged men, generous to a fault, but such strangers to compromise that the celebrated north-east poet J. C. Milne could write, 'Oh Lord look doon on Buchan and a' its fairmer chiels! [men], for there's nae in a' yer warld mair contermashious [argumentative] deils!' We were shortly to discover that Buchan's hard fist had also left its mark on the trout of the Ebrie Burn.

Little burns like the Ebrie are very sensitive to the cool rains of late spring. Still saturated after an especially cheerless winter, the sodden fields or parks could take no more on that April day long ago. Even as the showers cleared up, we knew that enough muddy water would have chuckled in from the russet red land drains to rule out fly fishing. Besides, the boys were still at that early stage in their fishing careers where the desire to catch the most fish regardless of method is in the forefront of the young sportsman's mind. Seeking to avoid unnecessary exposure to rat catcher's yellows (Weil's disease), we hunted the worms at the opposite end of the garden from the rhubarb crowns. There the previous summer, and to the horror of my dear wife, three friendly brown rats, one a little smaller than its companions, had made their temporary home under the broad green leaves, boldly emerging to slake their thirsts when a heavy shower at last broke the drought of 1976. It did not take long to fill the old honey jar with the red worms that we were soon to send tumbling into the Ebrie with two hooks and enough split shot to take them down to the ravenous trout that the boys' imaginations had placed at the foot of every little cataract and behind every glacier-worn boulder.

It was a mild morning but, in the minutes it took to put up the little rods and select the first and most enticing-looking worms from the honey jar, a gusty little breeze sprang up and with it a thin but soaking drizzle. First John and then Neil attempted to cast their baits into the tempting lies, but the rain had stopped the nylon line from running smoothly through the rod rings, and the precision of casting

underarm was impatiently abandoned in favour of frantic overhead thrashing, which served only to tangle the lines and catapult the worms onto the opposite bank of the tiny burn. More split shot and a brightening sky solved the problem and, time after time, snatching bites emptied the hooks. We still had some worms, but the best were now in the stomachs of the fish, so it was a great relief when at last a trout took a firm hold. But what sort of trout was it? Its sides flashed bright silver in the strengthening sunlight, and when at last

Jar of worms

John wound it in and swung it unceremoniously onto the sheep-cropped bank, we feasted our eyes on no ordinary fish.

Torpedo-shaped and almost a pound in weight, it possessed a brash glamour that outshone even the greatest prizes the boys had drawn from the home burn back at Craigdam. All of a sudden it came to life, the hook dropped from its soft mouth and only an immediate sweep of a fatherly gumboot foiled its attempted return to the Ebrie. All efforts to catch another came to nothing but, keen to avoid Neil going home empty-handed, I tied on smaller hooks and three or four of the worm-snatchers' careers ended in the polythene bag in which all of the catch was garnered regardless of size. By this stage, our supply of worms was low and we suddenly remembered that we were also responsible for Blidda's young life. Soon losing interest in the boys and their fishing, she had wandered down to the next of the burn's little pools. There she took off her right gumboot and, using it as a chalice, was seen drinking deeply from the Ebrie's murky waters. It was definitely time to get back to the battered VW Beetle, hope that Blidda's country-hardened immune system would cope with her newly ingested coliforms and admire the catch in the comforting fug of the kitchen.

The little trout seemed in all respects like the small brownies familiar to the boys from the home burn. Three were immature females but one, the solitary male and a little larger than the others, looked as though it might have spawned the previous autumn. As for the silver trophy fish, it turned out to be an immature female caught on its way back to the sea after over-wintering in fresh water. Years afterwards, I would learn how these very different trout were yet members of the same population, a population moreover whose structure had been moulded by the harsh environment of Buchan. With few trees or other tall vegetation to nourish them with wind-blown insects, yet with ample deposits of well-sorted gravel to shelter their eggs and early stages, life for the overcrowded trout of the Ythan catchment becomes one long struggle to achieve adulthood and thus

Immature sea trout

the opportunity to leave descendants. How it is that some of the males can do this without leaving the river while many of the females are required to risk their lives in the fertile but dangerous world of coastal seas was then a mystery to me.

One day, much later in my career as a fishery scientist and thanks to the devoted life's work of Andy Walker, my former colleague at the Freshwater Fisheries Laboratory at Pitlochry, I would learn the truth not just about the trout of the Ebrie but the whole family of primitive bony fishes of which trout are but one highly variable member. Both of us had been interested in fish from childhood. Over the years I had caught and kept in aquaria specimens of most of the game and coarse fishes to be found in southern England. But, having been brought up in Scotland with its much more restricted fish fauna, Andy's interest was more sharply focused on fish of the salmon family than was mine and it was correspondingly more penetrating. Where I had merely kept fish and watched them for short periods, Andy had bred them and grown them to adulthood in captivity.

Long before geneticists had demonstrated some of the important racial differences that exist between members of different trout populations, Andy had seen direct evidence of what appeared to be inherited differences in body shape, spot patterns and behaviour among his captives. He became especially interested in why it is that some trout migrate to sea and others do not and whether or not the tendency to migrate to sea is inherited. Certainly, the progeny of sea trout parents look a little different, even when reared to maturity in fresh water. Somehow their spots lack the vividness and definition of the brown trout in the local burns. At first we wondered whether this was merely the result of artificial feeding but, when Andy stocked a local hill lochan with them, he found that the dullness of the spots persisted for the whole of their lives and was still to be seen in the last known survivor of that generation, which weighed over seventeen pounds when it was finally taken on a stoat's tail fly and then driven smartly to the taxidermist's.

It was all very well noting the persistence of spot patterns but this observation told us nothing about the heritability or otherwise of the propensity to migrate to sea. The solution of the mystery lay not in the hatchery but among the trout populations of the Tay catchment. The fish leading the simplest lives were the merry sprites that inhabited the highest burns and were isolated from below by waterfalls that no trout or salmon had surmounted since the end of the last ice age some ten thousand years ago. Of course, the distant ancestors of the burn trout must initially have colonized the burn from the sea on the heels of the retreating ice sheets, but the long years of isolation, during which no emigrant was able to return to its birthplace to spawn, had virtually eradicated from the population any fish with a tendency to migrate. The result was a population of small trout perfectly adapted to the tiny burn its members never left.

A little farther down the catchment, where the burns were open to the main stem of the River Tay, the structure of the trout populations was more complex than it seemed. Sampled in the summer, the wee trout looked much the same as those in the burn above the falls, but go there after an October spate and you would see much larger hen fish cutting their redds in the small gravel of the spawning fords. A lesser number of cock fish of similar size would be in attendance together with numerous, burn-sized males and some burn-sized hens, all of them striving to help found another generation of trout. Andy was able to show that, after a year or two in the burn, the

Hen sea trout colouring up as the spawning season approaches

larger trout had migrated into the main stem of the river where the better feeding opportunities had enabled them to grow bigger and thereby lay more and larger eggs at spawning time back in the burn. In so doing, the migrants had to run the risk that they would be killed by predators such as pike, perch, big trout and saw-billed duck, which are more numerous in the main river. It is a risk well worth taking if you are a hen fish because supporting the growth of eggs requires more food than building up a supply of milt. For a cock fish, the main advantage of size is the capacity to intimidate rivals. Some take the risk of migrating but not as many, because the benefit on offer is not as great as that enjoyed by the hens. Needless to say, the trout themselves are unaware of the pros and cons at the time but are merely the long-term products of natural selection, which ruthlessly favours those that leave the most progeny.

In short, the tendency to migrate is even more strongly developed among the trout lower down the catchment where a considerable proportion, especially of the hen fish, migrate to sea before returning to spawn. Some of these fish may even see the coast of Denmark before they come home and in the process will grow almost as fast as salmon. Once again, there are pros and cons, more sharply drawn this time because, as we have seen, the sea is both rich and dangerous. Meanwhile, the link between the tendency to migrate and position in the river begs the question: how much of the spirit of adventure is inherited and how much is it the trout's response to its immediate surroundings? Andy was able to answer the question by following the fate of tagged fish transplanted from their sites of origin and releasing them at other sites up and down the river. What he found was that the tendency to migrate downstream in search of richer feeding is strongly inherited but that how far a fish actually travels depends on the point of release. Even so, he was able to generate a few sea trout from fish derived from sea-migratory parents released higher up the river than sea trout are found naturally and to generate a few more by releasing non-sea-migratory fish at the bottom of the river.

Around the same time, it was discovered that generous feeding early in the life of both trout and salmon triggers the earliest stages of puberty, which tends to suppress the complex hormonal and other changes that make migration to sea possible via the process called smolting. It could well be, therefore, that inherited differences in the tendency to migrate are mediated by inbuilt differences in the levels at which puberty and smolting are triggered. Wherever the whole truth lies, given the shrivelling bleakness of the Buchan landscape and the location of the Ebrie Burn low down in the catchment of the River Ythan, it is perhaps not too surprising that this grim little burn produces the odd sea trout.

ON LANCELETS, LAMPREYS
AND TELEOSTS

ONE of the things that the human brain seems to be rather good at is the recognition of patterns in the sights, sounds and shapes fed into it. Perhaps it is this facility that lies behind our sense of the aesthetic and why we take pleasure in music, mathematics and non-representational art. Certainly, seeing common structural patterns among animals is one of the many pleasures enjoyed by the serious naturalist, including those whose life's work has been among the fishes and their primitive relatives. The winding pathway that would one day lead me to look at salmon in this disciplined way, and so to speculate on its origins as a species, had its beginnings on a summer's day half a century ago.

Just as many a God-fearing Aberdeenshire farmer has been heard to complain of 'anither gran' hairst [harvest] day snappit up by the Sabbath', so the Great Headmaster in the Sky always seemed to me to reserve the sunniest of his days for the examination season. So it was that, on a blazing fifties morning, I stepped anxiously into the school's biological laboratory to face the practical examination, the results of which would either admit me to St Andrews University or consign me forever to some grim and colourless future, shuffling paper in an urban counting house. I am not sure what I expected to see laid out on the bench before me. My fear was that it would be a very dead dogfish, reeking strongly of formalin, its tail long since cut off by the

Lesser spotted dogfish

zealous myrmidons of Flatters & Garnett, the renowned old firm of biological suppliers, to reduce the costs of storage and carriage. Alive in its natural element, the lesser spotted dogfish, *Scyliorhinus caniculus* L., is a sinuously elegant animal, tweedily stylish in a speckled coat of shagreen mail, making a modest living snuffling for shellfish among the tide-swept sandbanks of the narrow seas. In attenuated death, sunken half-closed eyes giving no hint of the bright jewels that in life were its windows on the world, the dogfish of an A-level practical is a corpse indeed.

Another step through the door, and a nervous glance revealed no sign of any such sad mummy, neither did I see or detect the sneaky smells of any rat or frog in front of the low stools I and my fellow victims would shortly occupy. What, though, were the thin slivers of pink in the glass Petrie dishes? Surely we were not being examined on our ability to reveal and interpret the anatomy of earthworms? A peacock butterfly, fluttering in from the world of freedom that to us now seemed so grimly distant, temporarily distracted me. When I looked back, I realized with an inward eureka that what I thought might be earthworms were in truth lancelets, remarkable echoes of our earliest ancestry that I had read about but had never before seen in the flesh. Caught in the sparkling light of that distant midsummer morning, the herringbone pattern of the lancelet's coral-pink muscles appeared almost luminous under a transparent veil lit up with the rainbow colours of sunlight refracted by the walls of the single layer of cells that form its skin. All thoughts of the examination were swept away by that first sight of what used to be called *Amphioxus lanceolatus* (Pallas) but is now known as *Branchiostoma lanceolatum* (Pallas).

Zoologists love giving clumsy dog Latin names to the most beautiful of God's creatures even if, as with *Amphioxus*, the derivation is Greek and merely means pointed at both ends. As for *lanceolatus*, it is derived from *lancea*, the feminine Latin word for a light spear. So, a light spear pointed at both ends was not such a bad description of the softly shimmering shape I marvelled at on that bright morning. Whoever renamed the lancelet *Branchiostoma* must have had access not only to a Latin dictionary but also to enough magnification to see that its *stoma* or pharynx was perforated throughout its length by over a hundred branchiae, the gill-like slits that form the holes in the fishing net it uses to filter its microscopic prey from the sea water that gently caresses its sandy home.

My excitement at seeing my first lancelet was initially the purely sensual pleasure of seeing a shimmering object retaining much of the vibrant beauty it would have displayed in life, when all that I was expecting was a flaccid and mutilated cadaver that had long since lost any tangible link with its native element. Later, as I reverently took apart the fish-like creature in the Petrie dish, I travelled in my imagination to that remote era, over half a billion years ago, when the foundations of the group that would one day give rise to all back boned animals were being laid. Then, as with my lancelet, Cambrian fossils suggest that the segmented muscles of these hardy pioneers propelled their owners by contracting along either side of a springy compression strut called a notochord. It was not the most efficient way for a 'proto-fish' to put its swimming muscles to work and, in time, most members of the group evolved the segmented column of hard parts we call vertebrae to give strength and direction to their efforts. Fortunately for biological science, the primitive pre-vertebrate arrangement of an unreinforced notochord is still the best option for a small, shallow-burrowing animal that spends most of its time filtering sea water. 'If it ain't broke don't fix it' is as true of the effect of natural selection on the bodies of living things as it can be of conscious design when, as in the case of the traditional side-by-side shotgun or of the tea clipper, perfection has been achieved.

So successful is the lancelet design that its various species are found in shallow temperate and tropical seas all over the world. Without proper eyes, head or vertebrae, as many as five thousand individuals per square metre have been recorded in the shallow sand of Discovery Bay, Jamaica, and in parts of the Far East it forms the food of both people and domestic livestock. Only once have I seen a live one and it came from a sample taken in an anchor dredge, an ancient, heroic device that takes a great bite out of the sea bed. The late Norman Holme and I were surveying Mevagissey Bay in Cornwall from the Marine Biological Association's capable but rather lively research vessel *Sarsia*. As we tipped the contents of the dredge on to the sorting table, out leapt the startled lancelet, flipping stiffly from side to side just as the results of my schoolboy dissection suggested it would. I should have condemned the little sprite to a jar of formalin along with the rest of the biological sample, but instead made a note of its occurrence and returned it, still merrily vibrant, to the sea foaming alongside. I wonder if it made it back down to the sea bed. I would like to think that it did, but I fear that, not long after

Lancelet

leaving the surface, it would have found its final billet in the stomach of one of the many mackerel that, in those pre-mid-water trawling days, thronged the broad western throat of the English Channel.

Some years before my first surprise encounter with the lancelet in the Petrie dish, my brothers and cousin and I had spent many a summer day collecting fish from the River Chess, that chalk stream tributary of the middle Thames whose waters were usually so clear that to us it seemed like the perfect natural aquarium. The dappled miller's thumb, *Cottus gobio* L., under the flint cobble, the rust-coloured stone loach, *Barbatula barbatula* (L.), sheltering under the waving emerald fronds of the water crowfoot, the merry shoals of minnows, *Phoxinus phoxinus* (L.), sporting over the gravelly runs – here was a living masterpiece indeed, whatever the means of its creation. It was in just such an aquatic Eden that we first saw the pencil-thin Planer's lampreys, *Lampetra planeri* Bloch, lifting the flint pebbles with their jawless sucker mouths to build their shallow nests. Warm slate grey above and pearl grey below, never had they been more alive as their single nostrils drank in the intoxicating pheromones of their consorts and their perfectly circular eyes shone upon the bright pebbles and squirming bodies that filled their vision. Within weeks all would be dead; the tan-coloured denticles in sucker mouths that had never rasped living flesh would be the last of the limp remains to decay among the muddy roots of the watercress.

It was my resourceful cousin Stuart who was first to discover how the short-lived pearly lovers had spent the gentle years of their childhood. The frames of our nets were crudely made of soft wire from the ironmonger's, good enough for sampling the creatures that lived above the river bed but far too bendy to investigate what lay below. He had constructed a new net with a strong frame of fence wire. It was not strong enough to dig into the compacted pebbles in the riffles but more than adequate to dig deeply into the soft, silty mud that had built up near the bank. The first hauls were disappointing, mud and more grey mud relieved here and there by the flattened

My brother Peter in the River Chess

bodies of snail-eating leeches and the pale, burrowing nymphs of a species of mayfly. Then it was that he saw them, mud-coloured, eyeless slips of flesh with a sort of hood where the mouth should be. Most were no more than two or three inches long but a deeper thrust of the net brought forth a larger, more active one. It also appeared to be blind, but behind the hood what appeared to be the beginnings of eyes peered uncertainly through the translucent skin. There in the mud, Stuart had found the developing larvae – biologists call them 'ammocoetes' – of the lampreys we had watched lifting the stones with such apparent purpose. Not only that but he had captured one part way through the complex metamorphosis that would turn the passive larva into a lively adult.

All four of the boys on that distant summer morning would one day become professional biologists and it was only then that we would understand the full significance of what we had seen. The ammocoetes in their burrows were living on micro-organisms they had filtered from the water above the river bed, trapping them in mucus secreted by a structure called an 'endostyle'. In many ways their lives at this stage resembled that of a lancelet that had left the hurly burly of the wave-torn sea coast to seek a less demanding living where predatory jaws were fewer. The German contemporary of Charles Darwin, Ernst Haeckel, would have explained the ammocoete larva of the very different adult lamprey as evidence for his theory of 'Recapitulation', which stated that each successive stage in the development of an individual represents one of the adult stages in its evolutionary history. In other words, one of the ancestors of the lamprey was a filter-feeding burrower resembling a lancelet. We now know that the theory, one that Haeckel greatly overstated at the time, was seriously flawed. However, the 'chordates', the ancient group of which lancelets, lampreys, salmon and even ourselves are all members, is remarkably uniform in the main features of its body plan. The genetically driven changes in developmental emphasis, which ultimately lead in very different final directions, set off from a similar embryonic starting point in which many earlier developmental steps are retained. Given that it is generally believed that all backboned animals are derived from a lancelet-like ancestor, we ought not to be too surprised that one of the most primitive has a lancelet-like larva. However, lest we look too far down our noses at such humble creatures, we should not forget that, although we do not have an ammocoete larva, we do have an endostyle, only we call it a thyroid gland.

The Chess lampreys were unusual in that their metamorphosis was followed immediately by reproduction. Most lampreys grow to a larger size before breeding by attaching themselves to fish with their specialized sucker mouths and rasping their flesh with denticles hardened by tanned protein. The muddy shallows that provide the best feeding opportunities for the ammocoete lack fish large enough to accommodate the sterner needs of the adult, so usually the switch from filter feeding as a larva to a particularly grisly form of parasitism involves a downstream migration to a lake, an estuary or the sea. The immediate ancestors of the Chess lampreys must once have been migratory, swimming upstream to breed and establish the larvae in the shallows and downstream to grow to full size. When, as seems to have

Sea lamprey, lampern, Planer's lamprey and pride

happened in many river systems, upstream migration is inhibited by natural obstructions like waterfalls, there is strong selection pressure against downstream migration and in favour of reproduction without a further period of growth. Given that the Chess lay under many feet of ice as recently as ten thousand years ago, the evolution of today's Planer's lamprey from a migratory ancestor must have taken place very recently. Indeed, most specialists believe the ancestral species to have been the extant river lamprey, *Lampetra fluviatilis* L., a highly successful estuarine migrant with a life cycle that parallels that of the sea trout in much the same way that the life cycle of its much larger sea-migrant relative, *Petromyzon marinus* L., parallels that of the salmon. Given that there has been a succession of glaciations over the last few million years, it is quite likely that, as with the evolution of freshwater-resident brown trout populations from sea-migratory antecedents, the evolution of land-locked lamprey populations has happened more than once.

It was the distinguished Scottish genealogist, the late Sir Iain Moncrieffe of that Ilk, Albany Herald of Arms, who reminded his readers that no family is older than any other. It's just that some of them think they are because they can trace their ancestry further back than most of us. The same applies broadly to fishes, since the antecedents of some – the coelacanth, for instance, whose close relatives can be seen in strata three hundred million years old – are better represented in the fossil record than those of others. Fishes that live above muddy sediments stand the best chance of stony immortality and those that live in fast-flowing, gravelly rivers the worst. Salmon spend an important part of their lives, and die after spawning, in just such geologically unpromising places, so it is perhaps not too surprising that their fossil record is so sparse.

Unlike coelacanths, which are highly specialized lobe-finned fishes distantly related to our own amphibian ancestors, salmon are members of a diverse group of lightly-boned fishes that ichthyologists, the biologists who study fishes, call 'teleosts'. The wide separation between the two pairs of fins of salmon – they are homologous with our arms and legs – and the fact that its swim bladder is not closed off, as it is in the most advanced bony fishes, but opens into the pharynx like a lung, is evidence that the salmon is a comparatively primitive teleost. Teleost fishes first appear in the fossil record in rock strata laid down some two hundred million years ago, but the first salmon-like fossil, the curiously named *Eosalmo driftwoodensis* Wilson, is only some sixty million years old, by which time many much more advanced families of bony fishes were already well established. For years, it seemed that the evolutionary history of salmon would ever remain a mystery – not that the absence of objective evidence in any way prevented the bolder scientific spirits from stating their strongly held views. Fortunately, advances in the study of salmonid genes, and the ways in which they are organized into strings called chromosomes, have recently become available to help ichthyologists make better use of their knowledge of living and fossil members of this important family.

As in all animals, most of the DNA (deoxy-ribose nucleic acid) that makes up the genetic material of salmonid cells is held together by the internal membrane that bounds the nucleus of the cell. Additional genetic material is present outwith the nucleus in bacterium-like structures called mitochondria. Unlike the so-called nuclear DNA, which is carried on two sets of chromosomes, one from each parent, the mitochondrial DNA is passed down the generations only through the female line. It therefore remains unchanged as generations succeed one another. In other words, mitochondrial DNA behaves like a maternally-inherited surname. Just as with written surnames, 'spelling' mistakes sometimes take place between generations. These errors are very rare and are caused by small molecular inaccuracies during replication. Geneticists call them 'mutations'. It is possible to estimate the rate at which they occur and to use such estimates to guess how long ago past turning points in the inherited history of a species took place. Because rates of mutation are not constant,

Fossil teleost fish

however, they can never provide a foolproof measure of the passage of time. Experience with using the technique to follow the recent evolution of our own species suggests that it tends to underestimate the antiquity of past events such as our own ancestors' diaspora from their African calf country. The same may well apply to salmon and, for that reason, ichthyologists prefer to use mutation rate-based estimates in conjunction with other evidence, including what is known of the fossil record and of the structure and life cycles of the salmon's living relatives. Further clues to the relationships between salmonid species come from the ways in which the nuclear genes are organized into chromosomes and, within species, in the representation of particular versions of genes in different populations. Taking all these pieces of evidence together, the evolutionary history of the salmon and its relatives is currently believed to have taken the following pathway.

The group of fishes that ichthyologists call teleosts is something of a ragbag of only loosely related species. The main characteristic of most of these species is a lighter and more delicate bone structure than their Triassic (180- to 230-million-year-old) ancestors. The salmonids are members of a broad group of teleosts which include the herrings, lantern fishes and smelts and which store a good proportion of their fat reserves in their muscle tissue. Since the bulk of these fishes consists of muscle tissue, the effect of this adaptation is to distribute accumulated fat fairly evenly throughout the body so that the centre of buoyancy remains essentially unaltered both as the fat is built up during the growing season and later when it is mobilized to fuel spawning. It is an adaptation that is especially common in mid-water fishes feeding on plankton, which fluctuates widely in abundance, and it may help them to store large quantities of surplus energy quickly during the weeks of plenty.

Some of these fatty teleost fishes possess what biologists call an 'adipose dorsal fin'. It sits in front of the tail on the dorsal surface of the fish. Strictly speaking, it is not a fin at all because it consists largely of fatty skin and connective tissue and is unsupported by rays of bone. Rather like the giant panda's thumb, which is not a thumb at all but an additional digit that probably evolved in response to this animal's highly specialized dependence on bamboo shoots, the adipose dorsal fin seems to be the evolutionary response to some other selective force. Because similar structures have evolved independently in quite unrelated groups of fishes, the fin must have real survival value for its owners. What the fin is really for and exactly how it helps fishes to live long enough to breed and therefore perpetuate the adaptation was something of a mystery until recently. However, important clues to its likely function have come from the work of my former colleague, Dr Clem Wardle, lately of the Marine Laboratory in Aberdeen, who is an expert on how fish swim. He has found that at speeds up to three body lengths per second, the side-to-side movements of the posterior half of the fish generate little turbulence and consequently little drag.

As speed increases, so does turbulence, but not to the extent that simple calculations might predict. This is because of vortices shed from the fast-moving tips of the tail fin and of the anal fin just in front of it. The adipose dorsal fin has been separately evolved to fulfil the same function on the top surface of the fish, thereby enabling it to swim fast efficiently whenever the need to escape predators or to catch agile prey requires it.

But which of the relatively primitive, fatty teleost fishes might have given rise to the salmon family? Without access to a time machine, we can never be certain. However, cytogeneticists, those scientists who study the ways in which genes are organized into chromosomes, tell us that the earliest salmonid fishes had double the number of chromosomes of their immediate ancestors. These ancestors appear to have been members or close relatives of the smelt family whose current members include both fully marine and euryhaline – possessing the ability to live in both fresh and salt water – species. The mistake in reproductive cell division that led to the doubling up is thought to have taken place some one hundred million years ago and to have done so in one or more of the euryhaline types. Why this aberration persisted and what exactly happened next, we

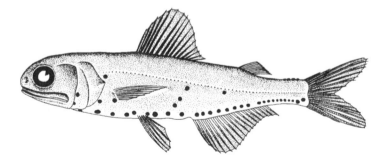

Lantern fish

have no direct means of knowing. One intriguing possibility is that two related species with different numbers of chromosomes had been hybridizing for some time, producing offspring that exhibited what breeders call 'hybrid vigour' and had coped especially well with the prevailing environmental conditions. However, even if they had thrived much better than either parent, they would have had problems breeding. Effective reproduction involves the pairing up of homologous chromosomes – the ones carrying the same sequences of genes – contributed by each parent. Pairing like this, therefore, requires that each parent has the same number of chromosomes, and that is why most hybrids are either infertile or have so low a fertility that their progeny rapidly die out. An 'accidental' doubling up of chromosome numbers in the reproductive cells of the hybrid, so that each parent contributes the equivalent of two sets of chromosomes to its progeny instead of one, has the potential to restore fertility provided that there are similarly fortunate individuals nearby with which to mate. It all sounds rather a chancy business but, as Charles Darwin showed, the effect of natural and sexual selection on chance events is the very engine of biological evolution for both salmon and men.

Possessing a double set of chromosomes is, on the face of it, rather a clumsy way for an organism to carry its genetic blueprint. As Emile Levassor, the pioneering motor-car manufacturer, said about one of his gearboxes, 'Ç'est brusque et brutale mais il marche'. So it is with many a biological characteristic. If its advantages outweigh its disadvantages and there is no more competitive alternative, the impersonal forces of natural selection will allow it to persist down the generations. One feature of polyploid – the name given to organisms with more than two sets of chromosomes – plants is that they tend to grow larger and faster than their less well-endowed relatives. Had the advent of the salmon family coincided with increased feeding opportunities, the additional genetic material may even have been advantageous. Migration from a freshwater or estuarine environment to the richer pastures of the open sea could have provided just such an opportunity. One of the disadvantages, though, of having a large number of chromosomes is that it increases the theoretical risk of 'mix-ups' when the two sets of newly replicated genetic material are pulled apart as cells divide. It is a problem that probably looms larger in the minds of cytogeneticists than in the dividing cells themselves. Nevertheless, it appears that over the last one hundred million years, there has been a tendency for salmonid chromosomes to join together, so reducing the complexity of cell division. Interestingly for the evolutionary detective, what seem to be the most archaic of the living salmonids tend to have the largest numbers of chromosomes and the most recently evolved have the fewest. On this basis, and in the flickering light shed by the study of the competitive anatomy and biochemistry of living and fossil species, it would seem that salmonid evolution may have proceeded as follows.

Some one hundred million years ago, a doubling up took place in the numbers of chromosomes of a smelt-like fish that might itself have been derived by hybridization between two closely related species. Perhaps it lived like the smelt or sparling of today, *Osmerus eperlanus* (L.), a small, dully silver fish that smells strongly of cucumber and looks rather like a miniature salmon. It breeds at the head of tide and, if cut off from the sea, is able, like the salmon, to grow to adulthood in fresh water. Normally, however, it makes most of its growth in shallow estuaries and coastal seas, places where stirred-up mud often reduces visibility and where a strong, species-specific scent may help shoal members to stay safely in touch with one another. Some seventy million years ago, the ancestors of the white fish and graylings separated from the salmon line, the latter having so marked a herbal scent that taxonomists know them as the sub-family *Thymallinae*. Next to branch off were the charrs and their relatives and probably also the various Pacific species of salmon and trout. The immediate common ancestor of our Atlantic salmon and sea and brown trout is believed to have been a coastal sea-migratory

Smelt

species breeding in rivers entering the north-east Atlantic Ocean and which is variously estimated by different geneticists to have split between two and ten million years ago. The trout never naturally reached North America – the American brook trout is a riverine species of charr – but the Atlantic salmon did, perhaps as recently as 600,000 years ago. Differences in numbers of chromosomes and in bodily proportions now distinguish the two varieties of Atlantic salmon, but they have yet to split into completely separate species.

Despite the paucity of its fossil record, one of the conclusions we can legitimately draw about the salmon and its living relatives is that they are simple variations on rather an ancient theme. It is almost as if, like the coelacanth, there were millions of years when nothing much changed and then, for some reason, the evolutionary process burst into life again. Perhaps the doubling up of chromosome numbers was the catalyst that enabled the early salmonids to exploit a variety of new habitats. Another factor is that when fish like salmon come home to spawn they do so with remarkable accuracy. The result is that different groups of fish tend to become isolated from one another and to be moulded by natural selection to suit local conditions, much as a stockbreeder might select for particular characteristics in his animals. It is a process that still operates today and that is why, even within the same river, some populations may differ in features like the rate at which they develop, in the time of year they return from the sea and in the number and size of their eggs. The need for salmon to be adapted to local conditions is why artificial stocking with non-native fish is often unsuccessful and may even do harm by contaminating the indigenous population with ill-adapted genetic material. Nor does local adaptation end in the river. The timing and routes of migration at sea for different populations also differ and have evolved to maximize survival and ultimate reproductive success; the characteristics that geneticists call 'lifetime fitness'. When changes take place in the freshwater and marine environments that affect the distribution of food and predators in time and space, so too must the salmon. Down the ages, the remarkable capacity of their populations to adapt to new conditions has been the key to the survival of the species. The most important and widespread of these natural changes are associated with changes in climate, the subject of the next chapter.

CHILDREN OF THE SUN

THE modest planet that bears us, its all too arrogant passengers, on a thin crust of tectonic plates, is doubly blue. Cloud-flecked blue skies above blue seas are the precious gifts that set our world apart from its fellow solar orbiters. Without an atmosphere swirling capriciously above the irresistible currents of the oceans, the earth's climate would be so extreme that even the humblest of micro-organisms would struggle to exist. Between them, air and water contain and redistribute the sun's energy. So effectively do they achieve this, that not only was the earth able to host life's anaerobic beginning but later it was able to support the development of bacteria capable of drawing the energy they required to grow and reproduce directly from sunlight. It was an evolutionary step that was to prove as important to the history of our planet as the origin of life itself.

The most important waste product of photosynthesis is oxygen. So successful were the new children of the sun that over time they produced so much of it that it transformed the composition of the atmosphere and with it that of the dissolved gases in the waters below. Combinations of bacteria held within a common membrane coped best with the new conditions, and it is from these, the first nucleated cells, that we ourselves would eventually evolve and the wealth of exquisite life forms that surround us. The ultimate atmospheric result was a new gaseous balance in which oxygen produced by photosynthetic organisms was used to liberate respiratory energy in both their own cells and those of the 'animals' that preyed upon them. The carbon dioxide and water so produced were then available to act as the raw material for more photosynthesis and hence more atmospheric oxygen. Carbon dioxide and water vapour are among what the popular press calls greenhouse gases, as is the methane produced when dead organic material decays. Along with other gases, aerosols and dusts caught up in the air, they absorb energy and reduce losses of radiant heat from the atmosphere. The balance struck between the capacity of photosynthetic organisms to deplete atmospheric greenhouse gases and the efforts of all organisms to add to them through respiration and putrefaction is a major intrinsic influence on world climate. It is by no means the only one. Volcanic activity is a potent, if intermittent, source of carbon and nitrogen oxides and of fine particulates. It is also a source of sulphur dioxide, whose presence in the atmosphere reflects solar energy and therefore tends to cool it.

Our understanding of climatic natural history is complicated further by differences in the amount and location of solar radiation reaching the earth's surface. Some of this variation is caused by changes in solar output, including the short-term changes associated with sun spot activity. The rest of the variation arises from the movements of the earth relative to those of the sun and which affect the location and seasonality of the arrival of solar energy at the surface, but not its total amount. The roughly cyclic changes caused in this way are popularly known as the Milankovich Cycles, after Milutin Milankovich, the Serbian mathematician and civil engineer who first described them in a quantitative way. The cycles have three components. The first of them is a direct result of the eccentricity of the earth's orbit around the sun. Because the orbit is not a perfect circle but is variously elliptical, there are times when the earth is closer to the sun than at others, so increasing the intensity of the solar radiation that reaches the surface of the earth. The relative eccentricity of the earth's orbit varies over a period of some hundred thousand years and is currently at the low end of its range. The second of the Milankovich Cycles arises from the fact that the axial tilt of the earth relative to the plane of its orbit – the persistent lean that gives rise to the seasons – varies from 21.5 to 24.5 degrees over a period of some forty-one thousand years. The greater the angle of tilt, the greater is the difference in the distribution of solar radiation between summer and winter and hence the difference between their average temperatures. By contrast, the differences in intensity between the solar radiation reaching polar latitudes and those reaching equatorial ones are greatest when the angle of tilt is smallest. The warmer winter air would hold more moisture and, provided the earth as a whole was not too hot, polar snowfall would tend to be greater. Less of it would melt during the cooler summers, and so it is possible that the growth of ice sheets is more likely at low angles of tilt. At present, the axial tilt is in the middle of its range at around 23.5 degrees.

Not only does the earth tilt, it also wobbles slowly as it rotates on its axis. This 'precession' gives rise to the third of the Milankovich Cycles, which has a periodicity of some twenty-three thousand years. It affects climate by altering the relationship between the times of midwinter and midsummer relative to the shape of the earth's orbit. The greatest seasonal contrasts arise when winter coincides with the greatest distance from the sun and vice versa. At present, midwinter nearly coincides with the shortest distance from the sun. Milankovich Cycles, vulcanism and biological and industrial activity all affect our climate in different ways in and over different time scales. We are currently undergoing a period of warming at a rate that alarms many, especially those who have short memories and do not fish for a living.

As the 1950s shaded into the 1960s, the ancient method of fishing called drift netting – 'Fairest way of fishing there is', according to my late shipmate, Charlie Button – was revolutionized by the advent of a new type of netting material. The material was monofilament nylon. Expensive and liable to tangle, it would have offered no advantage to Charlie and his fellow fishermen, pitting their wits against the

Sailing lugger drift netting for herring

teeming shoals of North Sea herring. They had in any case seen
their industry ruined by over-exploitation, including trawling on the
spawning grounds, during a period when the natural survival of post-
larval herring was lower than average. The North Sea fishery that
was to gain most from the new material was one of the smallest,
and it was prosecuted by a small group of hardy men fishing from
beautiful traditional boats called cobles off the bleak and rocky coast

of north-east England. Rarely out of sight of land, they had no one quarry but made a precarious living long-lining for cod, haddock and whiting and setting fleets of pots for crabs and lobsters. A few set drift nets, similar to but shorter and with larger meshes than the ones used by the herring men. They shot them in the gloaming and fished them during the brief hours of summer darkness. The fish they sought were salmon on the last leg of their homeward migration to the Coquet and mighty Tweed. They had to fish at night because otherwise the salmon would see the braided natural twine of the net and avoid it. The result was that they did not catch many and there were more than enough fish left to sustain the livings of both netsmen and anglers and to sustain the spawning populations on which the next generation of salmon depended.

All, however, was transformed by the new material. The salmon could not see the meshes so the nets could be fished night and day. Diving sea birds like guillemots and puffins could not see them either and neither could porpoises. The salmon catch increased, admittedly now of fish cruelly scarred round the middle by the thin, unyielding nylon twine, as did a grisly by-catch, discarded at sea, of birds and occasional sea mammals. It was not long before the wonderful salmon catches yielded by the monofilament nets came to the attention of Scottish inshore fishermen who had lost their herring and were catching too few haddock to make more than a subsistence living.

Haddock is a species whose breeding success can vary hugely, up to an hundredfold, and there had not been a really good year since the men had come home from the war. Soon, fish houses that had only seen shorter and shorter lines of boxes packed with the lank bodies of the various species of white fish, were enlivened by gleaming rows of plump salmon. Strong protests, especially from the owners of prosperous coastal net fisheries, that both economic and biological harm was being done provoked the government into appointing a committee, chaired by a judge, to look into the evidence. It was by no means clear-cut. All of the salmon fisheries, including one at West Greenland that had recently introduced nylon monofilament nets and others in Faeroese waters and the Norwegian Sea using long lines, were reporting good catches. The fact was that salmon resources across the North Atlantic were on the threshold of a period of high natural survival at sea that was to last into the early seventies. They were not the only northern species to do well. A series of unusually good breeding years for North Sea cod, haddock and whiting (all members of the family Gadidae) were christened the gadoid outburst by delighted fishery scientists. Plaice and even cockles did well during a period of cooling that came to be known among some climatologists as a mini ice age, and it was not long before some of the bolder spirits were weighing up the likelihood of a new glaciation. In the meantime, money was being poured into the construction of new trawlers. The resulting increases in fishing power saw to it that all too many of the increased numbers of young fish entering the gadoid fisheries were being caught before they had achieved a fraction of their growth potential and, especially in the case of cod, the opportunity to breed.

Now forty years later and the gadoid outburst is long over, cod fisheries in the North Sea and on the Grand Banks have collapsed in the face of overfishing, reduced catches of wild salmon have been overwhelmed in the marketplace by a glut of flabby farmed ones, and there is no more talk of mini ice ages. Now the fear is of global warming, warming moreover to which the additional greenhouse gases (including carbon dioxide, nitrous oxide, methane and chlorofluorocarbons) generated by an industrialized human population make so important a contribution that many believe, were Man to 'turn from his wickedness', the warming might be halted or even reversed. That the earth is currently getting warmer is indisputable; neither is there any doubt that we have never burned more fossil fuels, cut down more rainforests or kept more cattle. Nevertheless, the recent history of our fisheries is a reminder that we are also going through a period of unusually high climatic variability. Just as Thomas Huxley doubted in his day whether Man's harvest could ever deplete the stocks of fish in the sea, so some have questioned how much of the current warming trend is caused by us.

It is a legitimate doubt, given that there is credible evidence from ice cores and marine and terrestrial sediments that periods of very rapid warming have happened before during the Quaternary Period, the label that geologists apply to the last 1.7 million years. However, meteorologists have found that, only by adding anthropogenic influences to the known natural ones, can their mathematical models of world climate account for the temperature history of the last 150 years. Irrespective of Man's contribution, and whether or not there is a realistic possibility of reversing it by changing industrialized lifestyles, what we can be sure of is that the profligate burning of the millions of years of solar energy stored in fossil fuels will be looked upon by future generations as an exercise in bad housekeeping that should, in any case, have been curbed.

Male basking shark

In the meantime, we can be sure that warmer seas will offer different opportunities to its inhabitants, including its fishes, and not all of them will be for the worse. The occasional appearance of blue fin tuna in Norwegian fjords, the increase in the numbers of basking sharks off the Scottish coast and the northward spread of bass so that their fisheries now encompass the whole of mainland Britain are recent positive responses to warmer conditions. The steep decline in the abundance of cod in the North Sea is a salutary example of a negative response. The warmer water reduced but by no means eliminated the feeding opportunities for the youngest fish. Scientists from the Marine Laboratory in Aberdeen have recently calculated that, had the species not been overfished from at least the late eighties, it could, despite the warmer water, have sustained a spawning stock of adult fish of over a quarter of a million tons until at least the end of the twentieth century. The sad truth is that, by that time, less than 50,000 tons of mature fish were left to found the next generation, a quantity that history has shown to be entirely inadequate to sustain the stock. The Aberdeen scientists also investigated the widely held belief among fishermen that the cod had merely moved farther north in search of cooler conditions. Using tags that record the temperature of the water through which the fish had passed between the times of tagging and recapture, it became clear that North Sea cod do not move far during their lifetime. Furthermore, they often tolerate warm temperatures in the Southern Bight without making any attempt to move to cooler conditions in the northern North Sea. Plainly put, it was the effect of overfishing at a time when the supply of young cod was affected by the warming of the North Sea that caused the collapse of the stock.

The sad story of the cod is that part of our response to global warming should be sensitivity to the changed needs of the creatures affected by it. The fate of Atlantic salmon has been a patchily happier one. The discovery some forty years ago that, by rearing the fry and parr in tanks of fresh water and transferring the resulting smolts to sea cages, the entire life cycle could be completed in captivity gave rise to an industry that rapidly grew so large in Norway, Scotland and latterly Chile that the market for salmon was transformed. The wholesale price of salmon fell in real terms, thus enabling angling interests first to buy out and close most of the drift net and long line fisheries on the high seas and later to buy out and close many of the net fisheries in coastal home waters as well. As it happened, the buy-outs coincided with climatically driven reductions in the marine

North Sea cod

survival of wild salmon so the anglers did not always see the benefits for which they had been hoping. Fortunately, angling is not a very effective way of taking salmon and usually enough of them survived to found the next generation. Not all the populations were so lucky. The earliest-running salmon, many of which shared their summer feeding grounds with North American fish in the waters of West Greenland, were hardest hit by reduced marine survival and only now are they showing slight signs of recovery. Populations in proximity to the caged salmon also suffered from a variety of the problems inseparable from intensive farming, and in Norway a dangerous parasite native to the Baltic was introduced to some of its best rivers. In both Norway and Scotland, native salmon populations were put at risk from the effects on their genetic integrity of breeding with alien fish that had escaped from the cages.

It is not easy, against this complex background of potential influences, and in the face of a sustained increase in populations of the Atlantic grey seal, one of the salmon's most formidable predators, to know quite how much of the changes we have seen in the abundance and structure of their populations have been driven by what is popularly known as global warming. The latter's principal selective effect on salmon seems to favour those populations in which sexual maturation is triggered after a relatively short period at sea; they have therefore returned to their home rivers sooner and been less exposed to the higher levels of mortality that currently seem to be associated with the warming of the North Atlantic. However, even some of these fish have encountered problems. Fishermen have been reporting an unusually high proportion of thin, poorly conditioned grilse in their catches. Careful examination of affected examples by Professor Chris Todd of the University of St Andrews revealed higher than normal levels of astaxanthin in their tissues, revealing that they had once enjoyed normal health but had been obliged to run down their fat reserves while still at sea. The problem appears to be associated with the effect of raised temperatures in midwinter on the abundance of food for the grilse in the central Norwegian Sea.

We do not know what the climatic future holds. The warming looks set to continue. Furthermore, there is a great deal of inertia in the atmospheric and oceanic forces that, under the prevailing solar regime, control our climate. Even if we stopped burning fossil fuels tomorrow, the earth would continue to get warmer for the rest of our lives, if not of our grandchildren's. We should at least be thankful that the possibility, widely publicized by the news media, that climatically driven changes to the circulation of the deep sea would so affect the global conveyer belt of ocean currents that the British Isles would no longer enjoy the company of the North Atlantic Drift – often incorrectly called the Gulf Stream, a West Atlantic current – seems increasingly unlikely. Were that to happen we would, for the first time since the last ice age, shiver in the kind of subarctic temperatures to which our high latitude entitles us. We must, though, accept that the earth is fast getting warmer and ask what the world now requires of us. There is nothing new about the answer. It is what we should have been doing ever since William Blake wondered whether his Jerusalem could possibly be built among 'dark Satanic mills'. It is time to stop wasting fossil fuels, to control the growth of our own population and to manage the living resources of land and sea with love and sensitivity. Only then we can we afford to lie back and enjoy the sunshine until the next reduction in solar output cools the northern seas and fills the salmon fisher's days with joy.

OF SALMON AND SEA LORDS

I F global warming is just a contemporary reminder that sharp
fluctuations in climate have been a recurrent feature of the last
thousand years, we can at least be thankful that they all took place
during what meteorologists call an interglacial. The variations in
temperature have been sufficient to affect the abundance of salmon
and to make minor changes to the edges of its geographical range,
but not to deny them access to the great northern rivers of Europe
and North America. That, of course, is exactly what happens during
an ice age, the last one of which ended as recently as 10,000 years
ago. However, it was not the disaster for salmon that common
sense would predict. Rivers and sea areas far to the south, too hot to
support salmon during an interglacial, have the potential to become
prime salmon habitats during an ice age, and the genetic resources
that provide the opportunity for the effects of natural selection to
fine-tune the salmon populations of today would have served the
same purpose then. Recent advances in our understanding of salmon
genetics shed light on where the European fish might have been
when the earth last shivered and the men of Cro-Magnon hunted
mammoths on the fringes of the tundra. Once again, the clues lay
in patterns of variation in mitochondrial DNA, the small fraction of
maternally inherited genetic material that mutates at the relatively
high estimated rate of 2 per cent every million years. Without going
into too much of its technical detail, a large international study has
concluded that, during the last ice age, European salmon took refuge

in two main areas. Furthermore, during their isolation from one another, differences arose in the mitochondrial DNA variants that these two groups of fish carried. In particular, the refugial group that provided most of the maternal ancestors for the salmon nowadays living in the rivers that drain the Baltic Sea acquired, while isolated, a novel variant that became over time the main or only lineage present. The other refugial group, from which the stocks of most other European rivers appear to be derived, was probably centred in southern Europe during the last glaciation, mainly in what we know today as the Iberian Peninsula. Exactly where the main ancestors of the Baltic stocks took refuge is less clear. However, the most likely candidate is a glacial lake in what is now the southern North Sea. It cannot have been the Baltic Sea itself, because that area was entirely covered in ice until some twelve thousand years ago. One of the unexpected outcomes of the mitochondrial DNA analysis is the conclusion that, as the ice retreated, the refuge in where the southern North Sea now is also seems to have provided colonists for many Icelandic rivers. Isolation of this lineage group in such a glacial lake would have strongly selected against fish undergoing long migrations into the North Atlantic. It is no surprise, therefore, that Baltic salmon are nowadays confined to that sea nor that many of Iceland's salmon tend not to make the kind of long Atlantic migrations so characteristic of the fish that populated the Iberian refuge and nowadays make similarly long migrations from the rivers of the British Isles and continental Europe into northern waters.

Be all this as it may, one of the bonuses of spending long hours in northern waters in search of salmon is the curious sense of companionship to be derived from encounters with fellow mammals, some with larger brains – seven times in the case of the sperm whale – than any man. I took part in one such research cruise in 2005. We had left the Norwegian port of Bødo the day before and now, with the Lofoten Islands beyond the horizon to the south-east, the upswept bows of the Royal Norwegian Research Ship *Johan Hjort* lifted gently as she towed a surface trawl at four knots through the long swell. We were there to undertake a multi-species survey of that special region where, during the middle days of summer, Arctic and Atlantic waters embrace under a bright sun that never rests and biological production races ahead at a rate that would do credit to the richest of rainforests. All creatures were to be sampled and counted from the northern krill and amphipods in the plankton net (a technological marvel that took eight samples from as many depths) to the wealth of salmon and other surface and mid-water fishes that sport and fatten among them. High on the bridge, his practised eyes scanning the surface for the blows, flukes and bodies of great and lesser whales, was the tall, spare figure of the cetacean recorder. Gentle-mannered scion of a famous Norwegian whaling dynasty, he had personally grenade-harpooned twenty minke whales earlier in the year. His task, though, both now and throughout the cruise, was not to kill but to note down the species and position of every whale that crossed our path. As it happened, his penetrating gaze was not the first to be rewarded by a whale sighting. We could see from the sonar display on the bridge that herring were about and aimed the trawl to take a sample from a shoal. We were still hauling the net when, fine on the starboard quarter, a bright flash of reflected sunlight arrowed

Humpback whale

through the aft windows of the wheelhouse and into the eyes of the fishing mate. Rapidly grabbed binoculars revealed that we were not the only herring hunters out that day. The distant heliographic signal had come from the gleaming dorsal fin of a male grampus which, with the rest of his pod of pied killers, was tearing into the shoal of fleeing herring that our trawl had temporarily opened up.

The killers were to be the first of many whale sightings before we eventually secured alongside an old coal wharf in Svalbard. Sperm, humpback, minke, a gigantic fin whale, more killers and white-beaked dolphins all took their place on our cetacean expert's chart. At the time, all of us were too busy with our separate watch-keeping tasks to reflect much on what we really thought about these our fellow seafarers. True, I had enjoyed the perfectly cooked fillet of minke whale that my generous Norwegian host had served the day before we sailed and had admired the clever way the grampus and his friends had rounded up the herring. But it was only later back at home in highland Scotland that I began to think more deeply about how I personally regarded whales. Contrary to earlier zoological opinion, the cetaceans are not separately derived from primitive insectivores like ourselves but share a common ancestor with even-toed ungulates like cattle and pigs; small wonder, then, that my fillet steak tasted so familiar or that open-mouthed, head-to-head charges followed by a shoving contest form part of the male agonistic behaviour of both the sperm whale and the hippopotamus.

Long before such counter-intuitive insights were revealed by the analysis of nuclear DNA and a reappraisal of the paleontological evidence, great whales were the stuff of myth and nightmare. The term cetacean is, after all, derived from *ketos*, the Greek for sea monster. There were Middle Eastern stories of floating islands that came to frighteningly vigorous life only when the men who had landed on their smooth black shores set light to their cooking fires, and who has not heard of how Jonah survived three days and three nights in the belly of a 'giant fish', identified as a whale by no less a biblical authority than

Scrimshaw on a sperm whale's tooth showing
the death of its owner

Martin Luther himself? Literary products of a credulous age, such
tales entertained by evoking wonder mixed with a gentle measure of
horror. They were aimed at settled agrarian peoples who could afford
to spend a little time frightening one another with stories of creatures
they were never likely to encounter in real life. Quite different were the
whale myths of the traditional Inuit, hunter-gatherers who not only
encountered whales in their day-to-day lives but also valued them as
key quarry species on which their long-term survival depended. Like
most ancient subsistence hunting societies that could not afford to kill
too many of their prey lest that prey became incapable of reproducing,
the Inuit possessed a respect for their quarry that was built into their
culture by a web of stories in which the hunter regards the hunted with
quasi-religious awe. Even written down in English, such melodic song
lines still have the power to move, as in this chant from the south-west
coast of Alaska:

> Come, oh sea lord, chief of the waters. We are your friends!
> We wish you well. We bring you to a place to do you great
>   honour.

You are dying, but your death will not be forgotten.

We will strip your bones of flesh, but we will send them back to
the sea that you may live again, so fear not.

Let us lead you to the Kaniagmiut, people who admire you, great
lord of the ocean.

So long as such sentiments guided the whale hunter, the populations
of his quarry were safe. The rules changed when whale products
became the stuff of commerce, a transformation that human and
cetacean history owes to the Basques. The word 'harpoon', meaning
that cruelly barbed spear, is derived from *arpoi*, which means 'to take
quickly' in Euskara, the unique language of the Basques, a language
so foreign to other Europeans that one baffled cleric remarked, 'The
Basques speak among themselves in a tongue they say they understand
but I frankly do not believe it'. No doubt by the eleventh and twelfth

Whalers attacking with hand harpoons

centuries, when they dominated the trade in whale meat and oil, the Basques had become shrewd enough entrepreneurs to conduct their increasingly valuable business in the languages of their clients.

As demand for whale products grew, so, over the centuries, the numbers of the most readily captured right and bowhead whales declined, the Basque pioneers lost their pre-eminence and increasing numbers of courageous men from both Europe and North America ranged the oceans of the world to secure a share of the accumulated marine riches that the life of each great whale represented. The snag was that the rate at which the whales could grow and reproduce was greatly exceeded by the capacity of the ever more efficient whaling fleets to kill them. For a time, catch rates were maintained by voyaging farther and shifting exploitation to less desirable species. The invention of the cannon-fired grenade harpoon in 1861 raised the stakes still higher by extending exploitation to the largest and least buoyant of whales. The inevitable result of that critical development and later of the factory ship able to process its harvest at sea has been the severe and disturbingly sustained depletion of most species of great whale and the actual extinction of the Atlantic grey whale. Now, only the minke, smallest of the rorquals, exists in numbers sufficient to provide an exploitable surplus for the whaling nations still willing to profit from it. Theirs is a long story, initially of great heroism, all too often of ruthlessness in the face of appalling cruelty and, finally, of criminal neglect of the most basic principles of marine resource management.

Fortunately, the last four decades have seen a revolution in public attitudes to cetaceans, at least among the more enlightened of the western nations. It is a revolution for which the linked disciplines of neuro-physiology and ethology deserve due credit. The large brains whose complex circuitry enables cetaceans to both 'see' and communicate with sound appear, in some species, to confer a sense of self-recognition and even a capacity for compassion. Could some whales really inhabit a mental world that parallels our own? We do

Minke whale

not know but surely we can now be certain that inflicting unnecessary suffering on such sentient beings is no longer morally justifiable. Is there nevertheless still a case for taking a sustainable annual crop of abundant species like the minke whale? Looked at as a problem in single species population dynamics, the answer is undoubtedly yes. Furthermore, if we could be certain that every culled minke died humanely, eating its delicious meat is no more objectionable and a great deal healthier than enjoying bacon and sausages made from the processed remains of the domestic pig, its distant and highly intelligent relative. Regrettably, however, no such assurance can yet be given and all too many grenade-harpooned minke whales are hung over the ship's side to bleed to death or have to be finished off with a big game rifle. Have I enjoyed my last whale steak? I fear so and am bound to conclude that the lives of all cetaceans are far too valuable for cruel disposal. How lucky we are to enjoy their continued presence among us.

THE REAL MEANING OF LIFE

꧁ ─── ꧂

BOTH whales and salmon are advanced backboned animals whose populations are currently at risk of serious depletion only from ourselves, another member of the same remarkably uniform group of organisms. As Richard Dawkins, the evolutionary biologist, has pointed out, even such highly developed living things as bony fishes and mammals can credibly be interpreted not as ends in themselves but primarily as elaborate mechanisms for perpetuating the genes they carry. That such complex creatures as whales, salmon and people exist at all is an indication of the intensity of the competition for reproductive opportunity since the dawn of life on earth some five billion years ago. It is also an indication of the important role of sexual reproduction – which might originally have evolved as a way of correcting errors in the DNA of dividing cells – in speeding up evolution by making possible the combination of advantageous characteristics in the progeny of different individuals. Without this increase in the amount of useful variation, it is doubtful whether even five billion years would have been long enough for the forces of natural selection to produce anything approaching the diverse life forms of today. Because of its importance as the gateway to successful reproduction, it is inevitable that creatures like salmon compete to take part in sexual activity just as earlier in their lives they compete for food and space. Charles Darwin recognized that the result of such competition was a special case of natural selection that he called 'sexual selection' and that, over time, has often resulted in the

Balmoral Castle as seen from the north bank of the Dee

development of energy-diverting bodily characteristics and patterns of behaviour that can put the participants at increased risk. They nevertheless persist because, on average, those individuals possessing them leave more successful offspring than those that do not.

Sexual reproduction in salmon is a surprisingly complicated business, and my colleagues and I were to learn about it at first hand from watching spawning salmon in the Girnock Burn, a tiny tributary of the Aberdeenshire Dee that tumbles through the forest of Abergeldie. The bounds of the estate march with those of Birkhall and Balmoral and, ever since Prince Albert leased its wonderful pink-harled castle as a temporary highland base for his eldest son, Edward, Prince of Wales, Abergeldie has had sporting links with the royal family. Nowadays the castle has reverted to its traditional function as the seat of the Gordons of Abergeldie, a branch of a large and ancient family.

It was Sir John Gordon who first leased and later sold the hamlet of Balmoral and its parent estate to Queen Victoria, whereupon Prince Albert planned and oversaw the building of a new castle on the site of the old. Both the wildness of the surroundings and the direct and kindly way – that quality Scots folk call 'couthiness' – with which the local people received their Sovereign enchanted the Queen like nowhere else in her kingdom. Above all, it was a place where she could be herself and enjoy friendships that completely ignored the stuffy conventions of court life far to the south. One relatively recent story suffices to illustrate how strongly this happy tradition continues. Her Majesty Queen Elizabeth the Queen Mother was out walking on Abergeldie with her lady in waiting. What the weather forecast describes as a 'wintry shower' was in full chilling blast when, around the corner of the track, appeared the kenspeckle figure of one of the Abergeldie foresters, a particular favourite of Queen Elizabeth. 'Dreadful day, Robertson,' observed Her Majesty. 'C—in' awful, Ma'am, c—in' awful,' was his honest but respectful reply.

My own first visit to Abergeldie was on a very different day over a quarter of a century ago in early December. I had not long joined the Freshwater Fisheries Laboratory at Pitlochry and two of my new colleagues had kindly offered to show me one of its best-kept secrets. Some years before, the laboratory had built upstream and downstream fish traps on the Girnock Burn so that records could be kept of the relationship between the numbers of adult salmon migrating upstream to spawn and the numbers of their progeny that ultimately migrated downstream to the main stem of the River Dee and thence to the sea. It had not snowed for some days but the route from Pitlochry included the Devil's Elbow and we did not want the embarrassment of troubling the emergency services. We therefore threw our gear into the back of an ancient Land Rover and set off in a vehicle that we knew would not let us down, even if it did have the riding qualities of a tumbrel cart. We need not have worried; before an hour had passed, the skiers' car park with its scampering snow buntings was behind us and we were coasting past the white-painted suspension bridges and cairn-capped mounts that distinguish the Balmoral estate from its Invercauld neighbour. Soon we had crossed the Dee onto the narrow south Deeside road and, turning off it onto Abergeldie, were bumping along the very track where once Her late Majesty and Robertson had exchanged greetings.

We were on our way to the Hampshires' Bridge, a rather flimsy piece of military engineering constructed by a territorial company of the Royal Hampshire Regiment as an 'Aids to the civil community' exercise. The bridge had once enjoyed brief tabloid fame in an article provocatively entitled 'The Road to Nowhere' by courtesy of an editor concerned that the civil community duly aided might only have been royal grouse shooting parties anxious to avoid wetting their well-polished Hogg's brogues in the Girnock Burn. The local beaters and keepers took no such unsporting view and anyway, some years before had not their royal employer brusquely condemned the tabloid involved as 'a bloody awful newspaper'? Standing on the bridge on that distant December morning, the bright winter sun glinting off the ice crystals in the refrozen snow and marvelling

at the trusting faces of the stags looking back at us, I remembered that in one of my pockets I had a Victorian, leather-covered flask that I had filled before we left with an honest but well-aged whisky blend. Drawing it triumphantly out from among layers of tweed and pouring a 'wee suppie' of the golden liquid into the measure, I asserted that surely now, under such majestic surroundings, there could be no better occasion than this one for my new colleagues to

Flask

share a nip of 'the auld kirk'. My suggestion was met only with looks of horror. Could it possibly be that their new chief, about whom they knew so little, was really a prime candidate for the Betty Ford Clinic?

They were too polite to ask, and my fingers were far too numb to pour the nip back into the flask. Of one thing I was certain; there was no way I was going to waste the nip so, mixing it with crunchy snow, I downed it myself and did my best to steer the conversation toward science.

Partway across a ford just downstream of the bridge was a patch of gravel that appeared lighter in colour than the rest. Not only that but the stones at the upstream end looked larger than the others. We were peering down at a salmon redd. The gravel looked lighter because, a month and a half earlier, it had been turned over and was no longer cloaked by the thin layer of darkly pigmented algae that covered the sediments elsewhere. The larger cobbles at the head of the lighter patch marked where the final strokes of the hen salmon's broad tail had lifted the smaller stones that had been swept downstream by the strong current over the short distance necessary to cover and protect the eggs she had buried. Not until several generations of her progeny had returned to spawn below the road to nowhere would we learn the full story of what lay behind the construction of the redd below the bridge. By that time, the strong currents that were so helpful to the spawning salmon had done for the original version of the Hampshires' Bridge, and my colleagues had installed remotely controlled CCTV cameras lower down the burn so that we could enjoy a salmon's eye view of the whole of the spawning process.

The hen salmon that had cut the redd had entered the Dee many months before, months spent mainly in quiet repose in pools way below the point where the Girnock Burn joined the river. Every so often, usually on the tail of a spate, she would make her way further upstream to some new haven where in deep water close to the bed she could conserve the energy she needed to complete her life's work. Throughout the whole of her time as an adult in the river, the sex hormones that had first triggered her return from northern waters had so suppressed her appetite that she had lost all desire to feed. Even

Hen salmon after some time in the river

had she wished to, there would have been nowhere near enough food in the Dee to support thousands of large fish. That was why she had to spend so much of her time resting. Not all of the salmon that came in off the tide were hens. Cock fish, some even larger and earlier-running than she, shared her first resting places. However, there were not as many of them and they had not yet developed the distinctive kype that is so characteristic of male salmon at spawning time.

Just why it is that some salmon leave their food supply in the sea behind and then enter the river so long before they are due to spawn has yet to be satisfactorily explained. Tagging experiments have shown that these fish tend to be those that have undertaken the longest marine migrations, often as far as the North-west Atlantic, where they share the feeding grounds of Canadian and American salmon in the waters of West Greenland. Long migration is also the hallmark of their behaviour in the river in that many make for the cooler and more remote parts of the catchment. One explanation for the early-running habit is that it maximizes the number of opportunities the fish has to take advantage of the periods of high water when migration is safest. These times tend to be concentrated in spring and autumn, so a winter- or spring-running salmon has two main chances to reach its distant spawning ford instead of one. The snag with this argument is that few of the fish appear to take advantage of it but spend the long summer months in deep pools or, where available, lochs part way up the river. Furthermore, the earliest-running grilse, many but not all of which are cock fish, usually enter the river too late to take advantage of the last of the spring high water, despite the fact that they also are making for the distant spawning areas where the male grilse will compete with the larger, older cocks for the favours of the hens. As we have seen, a large male salmon typically takes pride of place beside his betrothed, smaller salmon and grilse mount usually unsuccessful challenges, and as many as several dozen mature male parr, fish only five or six inches long and which have achieved sexual maturity without migrating to sea, gather excitedly around the hen, risking the snapping jaws of the dominant cock in the process. Nor are these small adventurers the only members of the wedding party. Salmon eggs are rich in fat and protein, nutrients in short supply as the winter tightens its grip, and so, once spawning begins, there is rarely any shortage of immature parr and trout eager to snap up any eggs the hen fails to bury. All in all, a hen salmon's nuptials are a somewhat crowded affair.

Given the enormous disparity in size between the dominant sea-run male and the tiny mature parr, the contest is not as uneven as it looks. As the quivering displays of the sailor home from the sea bring the hen to arching orgasm, the parr jostling about her spurting vent are also stimulated to release their highly concentrated milt deep into the redd. The intimacy of its mixing with the newly shed eggs gives parr sperm a competitive advantage, so much so that the results of recent DNA fingerprinting suggest that over a third of salmon are fathered by parr. The least successful cocks appear to be the small sea-run fish, most of which have returned after but a short period of sea feeding. The only hope for these unfortunates is to meet up with fresh hen fish arriving at the spawning fords after the larger, dominant males are spent. Long after the hens have dropped back downstream, desperate cocks hang about the redds, fighting among themselves and moving from ford to ford in unconscious hope of passing their genes on to a new generation. For most the hope is a

Leaping grilse

forlorn one, but natural selection has seen to it that their lives are not entirely wasted. By the spring, exhaustion and a gasping death have overtaken most of them and nutrients from their decaying corpses make a small contribution to the biological production that sustains the next generation of their relatives.

Some weeks – or even months in the case of large rivers like the Tay – after the earliest fish have buried their eggs in the gravel, other groups of salmon will be spawning low down in the catchment. Most of these fish will be much later-running than the upper catchment ones and, as a result, they have had longer access to sea feeding and will therefore tend to be larger than fish of the same sea age that spawned at the top of the system. By this time, some of the grilse will be at least as large as the earlier-running salmon and consequently more will father young than their opposite numbers farther up the river. Furthermore, because circannual events like running a river tend to happen later in the calendar year in younger individuals than in older ones, grilse tend to form a higher proportion of lower catchment populations. The differences in age structure and run-timing between the various salmon populations in the river, including differences in the proportion of cock fish that mature as parr, are all evolutionarily stable products of natural selection. The differences persist because salmon home at spawning time and, down the generations, this self-isolating behaviour helps to prevent locally useful inherited characteristics from being swamped by less well-adapted ones carried by fish from elsewhere.

We can therefore be sure that homing behaviour has played an important part in the evolution of salmonid fishes. In North America, where the Atlantic salmon populations date only from some six hundred thousand years ago and where there is no competition from the several sea-migratory and freshwater-resident variants of the European trout, *Salmo trutta* L., salmon populations have evolved that follow replicates of most of these lifestyles as well as the ancestral one involving long oceanic migration. Even more interestingly, some of these populations occur together, but without interbreeding, in the same river systems. It is not even necessary to wait six hundred thousand years. Even in Loch Rannoch, a deep lake within a few miles of my former home in Perthshire that acquired its fish fauna only at the end of the last ice age some ten thousand years ago, there are three quite separate populations of another salmonid fish, the Arctic charr, *Salvelinus alpinus* L. One type lives in mid water on plankton and insects and has a fusiform shape; another with large eyes and a more truncated body lives on what it can

Arctic Charr

find on the bed of the loch; and a third morph preys on fish and consequently grows much larger than the other two kinds. Strictly speaking, these separate populations are not distinct species because, should they care to, and who is to deny that on rare occasions they do, they could interbreed and produce fertile offspring. However, such virtually separate locally adapted populations show how purely behavioural differences can begin the process by which new species are created. Recently, cytogeneticists have shown how the evolution of salmonid fishes seems to be reflected in the extent to which an ancestrally high number of chromosomes has been reduced by successive fusions. As we have seen, although such chromosomal joinings may well have started life as 'accidents' in cell division, they make the replication of the genetic material easier and might thus have been favoured by natural selection. Individuals in behaviourally isolated populations carrying such rearranged genetic material would have difficulty in producing fertile offspring through interbreeding with members of other populations. The ultimate result would be the splitting of two related populations into two distinct species. Behavioural differences followed by chromosomal rearrangements certainly appear to lie behind the evolution of the Atlantic salmon and the European trout from a trout-like ancestor. However, the split is relatively recent, and a low level of hybridization between the two species still occurs in the wild and more frequently when escaped hen salmon encounter large wild trout at spawning time. It seems that, in the latter instance, intensive cage culture blunts species-specific sexual behaviour in farmed salmon so that orgasm is triggered even when the quivering male partner belongs to another, admittedly closely related, species.

When I passed on this little-known fact to a newspaper reporter, it duly appeared on the front page of *The Scotsman*. Unfortunately, the sub-editor seemed to have trouble telling his cocks from his hens because the story appeared under the erroneous headline 'Love is blind for amorous old trout'. The fertility of trout-salmon hybrids is low because European Atlantic salmon have fifty-eight chromosomes and brown or sea trout eighty, but it is not zero. For instance, an unusually large, but otherwise normal, Tweed sea trout of over twenty-eight pounds in weight was recently found to contain some gene sequences usually found only in salmon. This is an interesting outcome for two reasons. First, the frequency of hybridization between wild trout and wild salmon is higher among Tweed populations than for any other British river. Secondly, the River Tweed is known for the fast growth and large average size of its sea trout, many of which migrate to the Southern Bight of the North Sea, some even going as far as the German Bight and Danish coast. Their life cycle is more salmon-like than that of other British sea trout; indeed, in many ways it resembles that of the special race of Atlantic salmon that inhabits the Baltic Sea. It is just possible that the residual presence, in at least some Tweed sea trout, of DNA derived from past matings with salmon has been favoured by natural selection. That genetic modification can be found in the natural world will no doubt be something of a surprise

Salmon trout

to its biologically illiterate opponents.

The evidence of recent and continuing evolutionary change among populations of Atlantic salmon and its close relatives begs the question: why is there only one species of salmon in the Atlantic when – if two species that occur only in Asia and steelhead trout, sea-migratory populations of rainbow trout that have a life cycle closely similar to that of Atlantic salmon, are included – there are eight in the North Pacific? Part of the answer is that evolutionary change needs time. The Atlantic salmon may only have evolved a few million years ago, but since its emergence, a succession of ice ages and interglacial periods has shifted its range in a north–south shuttle often enough to blunt the selective effects of local differences in its freshwater and marine environments. The genus *Oncorhynchus*, to which all Pacific salmon belong, seems to have arisen at least twenty million years ago from salmonid populations living in large rivers offering a wider range of growth and survival opportunities than the generally smaller and less diverse systems of Europe. Although the rivers and estuaries of the North Pacific basin were also affected by the advance and retreat of ice sheets, it would seem that sufficient ice-free water was available to support at least the core of the evolving species of *Oncorhynchus*. Thus not only was there the time required for greater diversity to emerge but larger differences in local conditions strengthened the selective forces that gave rise to it.

Although the various salmon of today are relatively recent products of evolution, the ancient group of bony fishes to which they belong has a long and in all probability convoluted history. Part of their success is due to the fact that they are able to control their buoyancy and thereby save energy, with the help of a gas-filled sac called a swim bladder. It is believed to be derived from a lung, an adaptation evolved in some distant ancestor living in organically enriched fresh water, a medium where oxygen is often in short supply. It is just possible that the estuarine smelts, from which the salmonids appear to be derived, represent an early stage in the process by which a freshwater fish began to exploit the resources of the sea. Evidence in favour of this idea is that some smelts are able to complete the whole of their life cycle in fresh water. However, it is at least as likely that the smelts are specialized inshore members of a group whose immediate ancestors were fully marine.

Perhaps one day the palaeontologists and geneticists will tell us. What we can be sure of is that the ancestors of the whales, which so often accompany us on our salmon surveys, evolved from river-dwelling land mammals some fifty million years ago, a process that took only some twelve million years. Given that the wonderfully successful group of backboned animals that zoologists call the subphylum Vertebrata is one of the most uniform in the animal kingdom, it might have been thought that aquatic adaptations like gills and a side-sweeping tail, devices that had served the fishy ancestors of all mammals so well, would somehow reassert themselves. That, of course, is not how evolution works. Natural selection can act only on the characteristics that are present at the time, not on what might have been present in the past. That is why, despite their ancient legal status as 'Royal Fish', whales breathe with their lungs and swim by moving their tail flukes up and down, a less efficient form of locomotion derived directly from galloping on land. It is also why they have to secure their oxygen supply from the air and require a greater relative volume of blood than other mammals (10-15 per cent of their body weight compared with 7 per cent in Man), and high concentrations of the intra-muscular pigment myoglobin in which to store the oxygen. Such adaptations are expensive in terms of the energy required to sustain them and may well be one reason for the low reproductive rate of whales, a characteristic they share with many diving sea birds and one that, over the years, has rendered their populations highly vulnerable to commercial overexploitation, even to the point of extinction in the case of the Atlantic grey whale. However, their warm-blooded independence from sea temperature, another mammalian characteristic, also enables them to be fully active even in the coldest water and has been demonstrated to confer a capacity for rapid migration and, in some species, an ability to deep-dive and return to the surface unmatched by even such advanced fishes as the blue fin tuna.

Salmon, whales and Man are fellow vertebrates with nervous systems flexible enough for each species in its own way to improve its chances of survival by learning from experience. Thus salmon can recognize and remember their kin, their home range and the olfactory memories that include those that guide the returning adults to their natal stream. Whales, of course, are capable of much more, including the ability to 'see' with sound and often to behave in ways that give every appearance of that consciousness that at one time was believed to be unique to man. So we have three kinds of animal leading very different lives and whose evolutionary pathways have followed very different routes, and can we really say that any one is biologically superior to the other? It would seem not because, looked at through

Grey whale

the eyes of the population geneticist, salmon, whales and Man are merely highly successful machines honed by natural selection to propagate particular sequences of DNA. In that sense, all three are equal with one another and indeed with all other species of organism living at the present day, including the most pathogenic of bacteria. Is this austere philosophy the ultimate meaning of life on earth, or is it merely a summary of the ruthless and impersonal mechanisms that have used chemical and solar energy to exploit the chemistry of carbon to its limits and in the process created biological riches that, uniquely among mammals, we have been given the central nervous equipment to appreciate? Ruthless and impersonal mechanisms perhaps, but creators of products so wonderful that it is as if the quality of divinity is built into the very fabric of the universe, the existence of which is itself the greatest of miracles. Even if we can no longer believe with Frank Buckland that we and our fellow creatures are the individual handiwork of a celestial master craftsman outwith the material sphere, but the result of divine processes integral to it, we can still share with him his profound love of the natural world in all its glory and the special place he ever accorded to his fellow man and to that iconic fish, the Atlantic salmon.

THE PILGRIMS

THAT last autumn at home had been a happy time. The first frost that had brought the great fish up from the pools to breed among the stones had also brought the stray eggs that were the parr's last good meal of the season. The happiness was not to last. The hens slipped back to the main river as soon as their redd-building was complete. Now only the frightening cock fish remained, slashing at one another with the kypes that set their mouths into a permanent snarl. Soon they too would be gone, the lucky ones to join the exhausted hens down in the river and the others to weaken and die in the burn, their slack bodies ravaged by fungi and bacteria.

It was time for the parr to leave the horrors behind, to wriggle under the thin ice and down the tumbling burn with the little fish she grew up with to seek the security of the Royal Dee itself. Somehow she felt different about her tiny companions. Earlier that year, she would have had no hesitation in attacking them fiercely, had any of them dared to challenge her choice of feeding station. Now they were her comrades, fellow pilgrims setting off together to face the greatest adventure of their lives. She still had her parr marks but the merest hint of silver also suffused her plump sides as the tight little shoal surmounted the last of the burn's rapids and she saw the full grandeur of the river for the first time. Only the bankside alders seemed familiar and, as the last of their shrivelled leaves kissed the surface above them, the fish stole downstream.

The parr had seen an otter once before, its sharp, white teeth tearing into the egg-swollen belly of one of last year's spawners. Then it had ignored her, content to allow a sprat to pass when there was bigger game on offer. This one was different. It was standing on a boulder when she first glimpsed it, marking its territory with a spraint. The otter saw the shoal in the same instant, its attention drawn by a flash of silver as one of the parr turned to avoid a sunken log. Sliding with sinuous grace into the river, it made silently for the shoal at a speed that no parr could match but, turning together, they were able to outmanoeuvre their assailant, which, just for a moment, lost sight of them. They were not yet safe. The otter's first rush had taken it away from the bank and for a few seconds he was confused. But his sensitive whiskers felt the pulsing vibrations set up by the whirring bodies of the terrified parr and, sweeping round to face the shoal, he fell among them in a confusion of splashes and leaping bodies. One unfortunate, momentarily disorientated by contact with a paw, lagged far enough behind the others to suffer an agonizing but mercifully rapid death clamped in the otter's tight-shut jaws.

The fleeing parr lost themselves among the alder roots, venturing out into the flow only after their heart rates had returned to normal and their heavily exercised muscles had recovered their strength. That winter, the alder roots and the rocks and pebbles caught up among them were to prove good friends to the little parr and her companions. Not only did they offer protection against otter, pike and goosander, but this was one of the best places to be when the softly peaty waters of the Dee became a tearing, silt-clouded torrent.

A sense of unease as the barometer fell was the parr's first indication that something awesome was on the way. She had met winter spates before in the burn of her calf country. Then she had coped by burrowing under the stones in unconscious imitation of her early weeks as an alevin. Little by little the river rose; twigs and small branches swept past and presently the body of an old hind that had finally lost its struggle to survive among the peat hags of

Braeriach. The usual humming hiss of the river at peace had now become a continuous, thundering roar, its grim monotony relieved by the sound of tumbling boulders. Battened down as she was under the bank, the safety of the parr would depend upon how well she had chosen her winter fastness. Sustained by the last of the stray eggs she had eaten, her body was now in a state almost akin to hibernation and she was not always fully aware of the hydraulic drama being played out above her. If the bank gave out now, she and her fellows would be swept down to the estuary long before they would be able to survive the dessicating effects of salt water. Hold out, though, the parr did and, as the pink-tinged rays of the pale sun of late winter played among the thin, knobbly branches of the alders, the river level fell and she felt stirrings that led her out from her shelter to join the rest of the shoal in the river.

The otter

There was something different about her now. Her body had become longer and sleeker and her fins blacker; as for her parr markings, they were barely visible under her new coat of silver. There were more of her kind to be seen too, fish that had elected to over-winter in the burn rather than risk the hurly burly of the main river. She was part of a great gathering of smolts that together would meet at the head of tide, feeding hard on the first of the year's insects until a May spate would take them all under the Victoria Bridge, past the old lighthouse at Footdee and out into the North Sea. Pilgrims indeed, their great adventure had begun.

Victoria Bridge

SOME WORDS OF THANKS

The idea that it might be possible to write a book about life's meaning around the life cycle of the Atlantic salmon was not mine but that of my patient editor, Angus Mackinnon. It is, therefore, to him, to Sarah Norman and the ever helpful team at Atlantic Books under the leadership of Toby Mundy that I owe the greatest debt.

However, but for my friend, Redmond O'Hanlon, explorer extraordinary and natural history editor of the *Times Literary Supplement*, my modest career as an author would not have begun at all. By inviting me to contribute material to the *Times Literary Supplement* on subjects as diverse as whales, oceanographic voyages and climate change, and which in places see the light of day again here, he lit a fire that glows fitfully yet. To him and the Editorial Team of the *TLS* I give my heartfelt thanks.

Most of the illustrations in this book are drawn from family photographs and books in the family library. A shining exception, for which I am enormously grateful, is a wonderful study in oils by my artist daughter-in-law, Patricia Shelton. Her painting of the salmon incised on the Pictish symbol stone at Glamis in Angus now hangs in a place of honour at home.

Finally, those who know me will be rather surprised to hear that I did not have to call upon my lovely wife to interpret a scrawled manuscript, but typed it myself! Thanks to Freda are nevertheless due for the support she has given me over more than four decades, not least during the writing of this book.

Richard Shelton
2009

INDEX

Aberdeen, 53, 55, 59, 61–2, 67, 69
  Marine Laboratory, 160, 174
  Victoria Bridge, 204–5
Aberdeen Press and Journal, 53
Aberdeenshire, 50, 58, 141, 149
Abergeldie, 6, 186–8
acclimatization, 115–17
acetamide, 35
Acre, 30
Act of Union (1707), 23
Adam, Alexander, 58
adipose dorsal fin, 160–1
adrenal hormones, 8–9
adventure schools, 57–8
Africa, 6, 159
Agassiz, Louis, 35, 118
Agricola, 22
Alaska, 181
Albert, Prince, 186–7
Aldrich, Lieutenant Pelham, 88
alevins, 46–7, 202
algae, 108, 189
algal blooms, 107
Almondbank, 36
Altries, 25, 59, 61–2
American brook trout, 164
ammocoetes, 155–6
amphipods, 98, 108, 179
Angus, 22, 43
Anning, Mary, 33
Arctic terns, 94
Armstrong, John, 75

Asia, 196
astaxanthin, 109, 176
Atlantic Fleet mutiny, 93
Atlantic Ocean
  continental shelf, 134–5
  gyres, 135
  hydrothermal vents, 91
  increasing width, 77, 90
  mid-Atlantic ridge, 88, 90
  Slope Current, 134
  subarctic mixing zones, 40, 107, 128
  Vøering Plateau, 135
  warming of, 175
  Wyville Thomson ridge, 81, 83, 134
Australia, 87
Avon, River, 24

bacteria, 14, 16, 108, 167, 199
  cyanobacteria, 105
  sulphur bacteria, 91, 106
Badminton Magazine, 27
Ballater, 53
Balmoral, 186–8
Baltic Sea, 175, 178, 195
Banchory, 62
barometer, patent, 81
barometric pressure, 36
Barra, 134
Basques, 182–3
bass, 101, 173
Bay of Fundy, 122
Beaton, John, 132
beavers, 31
Beeching, Dr Richard, 53

bees, 73
beetles, 112
  larvae, 47
Bergen, 131
Bett, Andrew, 36
Bible stories, 138, 180
Billingsgate, 43
Birkhall, 186
Blake, William, 176
Bland, Thomas, 59
Bødo, 178
Bompas, George, 33
Book of Common Prayer, The, 78
botulism, 69
Braeriach, 204
Braes of Gight, 141
brain
  human, 149
  cetacean, 178, 183
Breadalbane, Marquess of, 116
British Association, 116
British Empire, 38, 116–17
Brunel, Isambard Kingdom, 35
Buccleuch, Dukes of, 118
Buchan, 141, 143, 147
Buckland, Francis (Frank) Trevelyan, 24, 28–42, 46, 113–26, 199
  and acclimatization, 115–17
  and arithmetic, 37
  and fish culture, 117–24
  and fish passes, 40–2
  and horse meat, 113–14

and rats, 28–31
salary and expenses, 36–7
Buckland, Very Rev. William,
    32-3, 35, 114, 118
Burgess, Dr Geoffrey, 37–9
Burrishoole Fishery, 50
Button, Charlie, 169
Byron, Lord, 141

Caithness, 4
calcium, 106
camels, 109
capybara, 116
carbon dioxide, 168, 172
carnivores, 108, 112
carotenoid pigments, 12,
    108–9
carp, 118, 125
Carpenter, William, 81, 83
Carstairs, Elsie, 54
cats, 70, 74, 112
cell membranes, 64–5, 70
Chantrey, Sir Charles, 33
charr, 163–4, 193–4
Chess, River, 43, 98, 153–4,
    156–7
Chile, 174
Chiltern Hills, 43
China, 30, 88
chlorofluorocarbons, 172
chlorophyll, 106
Cholmondeley-Pennell, H.,
    60
chordates, 155
Christ Church, Oxford, 33
Christie, Catherine, 58
chromosomes, 158, 161–4,
    194–5
Church of Scotland, 99, 138
Clyde, River, 124
cobles, 43, 170
coccolithophores, 106
cockles, 172
cockroaches, 71

cod, 3, 171–5
coelacanths, 157, 164
conifers, 61
Coquet, River, 171
cormorants, 48, 94–5
Corn Laws, 111
Cornwall, 30, 152
Cowie, River, 6
crabs, 73, 108, 171
Craigdam Kirk, 139, 143
crocodiles, 35, 113
Cro-Magnon Man, 23, 177
Cromarty Firth, 93, 96
Cruden Bay, 101–2
Curiosities of Natural History
    (Frank Buckland), 28,
    31–2
Cushing, David, 48
cyanide, 44

Dalton, John Neale, 138
Darwin, Charles, 80–1, 87,
    155, 162, 185
David, King, 138
Dawkins, Richard, 185
Deal, 84
Dee, River, 5–6, 25, 53–5,
    57, 59, 61, 186–90, 201–5
Dee Salmon Fishing
    Improvement Association,
    62
Descent of Man, The
    (Darwin), 81
Devil's Elbow, 187
diatoms, 105–6
dietary innovations, 111–14
Dinah (dog), 109
dippers, 47
Discovery Bay, Jamaica, 152
Disraeli, Benjamin, 46

dogfish, 149–50
dogs, 35, 74–5, 112
dolphins, 4, 61, 180

Dom Pinchon, 118
donkeys, 35, 113
Dordogne, River, 23
Dorsetshire lias, 33
dragonfly, 47
Druie, River, 44
Drumlanrig Castle, 118
Drumtochty Castle, 58–9
ducks, saw-billed, 146

earth
    axial tilt, 169
    magnetic field, 131, 135
    orbit around sun, 168–9
    precession, 169
Ebrie Burn, 141–4, 147
echidna, 87
Edinburgh, 136
Edward, Prince of Wales
    (later King Edward VII),
    123, 186
eels, 125
eggs (ova), 14–15, 46, 109,
    191, 204
    and artificial culture, 118,
    122
    numbers and size, 120, 165
    survival rates, 50–1
    trout eggs, 146
Egypt, 31
eland, 116
Elisha, 138
Elizabeth, Queen, the Queen
    Mother, 187
Ellon, 141
endostyle, 155–6
England, 24
English Channel, 153
Eosalmo driftwoodensis
    Wilson, 158
Eton College, 38–9
Faeroe Islands, 172
Faeroe–Shetland Channel,
    81, 126, 132, 134

Faraday, Michael, 33, 35
Fasque, 58
Ferguson, Rev. Ron, 138
Field, The, 116, 122
Fife, 28
First World War, 93
fish culture, 117–24
fish passes, 40, 41
fish spears, 44
fish traps, 187
fishing, 169–76
  angling, 174–5
  drift net, 23, 169–72, 174
  long line, 23, 172, 174
  net and coble, 43
  sweep net, 20
fishing flies, 59–60
fishing rights, 23
fishing rods, 59–61
Fitzroy, Captain (later Admiral), Robert, 81
fjords, 81, 173
Forbes, Edward, 81, 83, 87
Forrest, Catherine, 51, 53–5, 57
Forth–Clyde valley, 124
fossil fuels, 172–3, 176
fossil record, 157, 159, 164
France, 41
Fraserburgh, 102
fry, 47–8, 50, 174

gadoid outburst, 172
Geneva, 137
George V, King, 138
Germany, 41
Gibbon, Lewis Grassic, 57
gill arches and slits, 76
Girdle Ness, 6
Girnock Burn, 5, 186–9
glaciations, 157, 172, 177–8, 196–7
Gladstone family, 58

Glamis symbol stone, 21–3
Glasgow, 29–30
Glen Dye, 58
global warming, 172–7
Glubb Pasha, 31
goosanders, 48–9, 202
Gordon of Abergeldie family, 186–7
Grampian foothills, 9, 19
Grand Banks, 172
Gray, Dr J. E., 116
grayling, 163
Great Depression, 93
Great North of Scotland Railway, 53
Great Southern Ice Barrier, 83
greenheart, 60
greenhouse gases, 168, 172
Greenland, 136
  see also West Greenland
grilse, 10, 13, 16, 128–9, 176, 190–2
guanine, 66
guillemots, 171
Gulf Stream, 176
guns, 59

haddock, 46, 51, 171–2
Haeckel, Ernst, 155
hagfish, 87
Hampshire, 24
Hampshires' Bridge, 188–9
Hardy family, 139–40, 143
hares, 31
harpoons, 182–3
Harris, 132
hatchet fish, 8
Hay, David, 50
herons, 5
herring, 3, 111, 160, 170–1, 179–80
Herschel, Sir John, 33
Highland Railway, 53

hippopotamus, 180
HMS *Beagle*, 80–1
HMS *Carmen*, 31
HMS *Challenger* expedition, 77–9, 83–91, 105
HMS *Hydra*, 140
HMS *Lightning*, 81
HMS *Modesty*, 88
HMS *Porcupine*, 83
HMS *Rattlesnake*, 83
HMS *Shearwater*, 83
Holden, Alan, 44
Holme, Norman, 152
Holst, Dr Jens Christian, 131, 133, 136
Holt, Vincent, 111–13
Hong Kong, 88
horse meat, 113–14
House of Lords, 38
Huxley, Thomas Henry, 83, 173
hybrid vigour, 162

Iberian Peninsula, 178
ice ages, *see* glaciations
ice cores, 173
ice sheets, 169, 197
icebergs, 88
Iceland, 135, 178
immune system, 14–15
Industrial Revolution, 23–4, 40, 46, 111
Inuit, 181
Invercauld, 188
Invergordon, 93, 98–9
Iona community, 138
Ireland, 50, 120
Islington Dog Show, 123

jellyfish, 94, 97
Johan Hjort, 133, 178
Johnson, Dr Samuel, 105
Jonah and the whale, 180
Jones, Dr Jack, 118

kangaroos, 116
Keay, Stephen, 36
kelts, 15–16, 36, 43, 61
Kelvin, River, 124
Kincardineshire, 57, 59
kingfishers, 47
Kinloch family, 59
krill, 3, 108, 179

lactic acid, 9
lamp shells (brachiopods), 87
lampreys, 43, 153, 155–7
lancelets, 150–3, 155–6
Langham Hotel, 113
lantern fish, 3, 95, 160–1
learned behaviour, 72–4
leeches, 155
lemmings, 28
Lerwick, 132
Levassor, Emile, 162
Lewis, 132
lice, 7
'lifetime fitness', 165
Lisbon, 87
Liskeard, 30
Living Races of Mankind, 27
lobsters, 72–4, 108, 171
Loch Rannoch, 193
Lofoten Islands, 3, 178
Logie, 5
London, 27–8, 43, 122
Louis XIV, King, 35
Lowe, James, 116
Lowestoft Laboratory, 131
Lunan Bay, 4
Lundy, 28
lungs, 197–8
Luther, Martin, 181
Lyell, Sir Charles, 33

Macdonald, Ramsay, 93
mackerel, 153
MacLean, Julian, 126
MacLeod, Nicky, 75

Magna Carta, 23
magnesium, 106
magnetite, 21, 131, 135
mammoths, 33
Marine Biological
    Association, 152
Maryculter, 59
Marykirk, 57
Matkin, Joseph, 88, 91
mayflies, 98, 155
Mearns, the, 57
Mediterranean Sea, 81
Medway, River, 23
Melbourne, 88
mergansers, 48
mermaids, 35
meteorology, 81
methane, 168, 172
Mevagissey Bay, 152
mice, 35, 113
midges, 19, 21
Milankovich Cycles, 168–9
miller's thumb, 153
Milltimber, 53, 55
Milne, J. C., 141
milt, 14, 118, 191
Minch, the, 132, 134
minnows, 153
mitochondrial DNA, 158, 177–8
Moncrieffe of that Ilk, Sir Iain, 157
monofilament nylon, 169, 171–2
Mons Graupius, battle of, 22
Montrose, 43
Moore, Dr Andy, 131
moose, 116
Moray Firth, 4, 94, 96, 98, 134
Morris, Desmond, 70
Murray, John, 90
myoglobin, 198

Napoleon Bonaparte, 30
Nares, Captain George, 78, 86–9
Naval Prayer, 78–9
nerve cells, 70–2
New Guinea, 91
New Zealand, 88, 122
Newburgh, 141
nitrates, 106
nitrogen oxides, 168, 172
North America, 110, 122, 164, 177, 193
North Atlantic Drift, 176
North Atlantic Salmon
    Conservation
    Organization, 136
North Esk, River, 5, 43, 99, 101
North Sea, 4, 15, 43, 101, 130–1, 139, 141, 178, 205
    fisheries, 170–4
    German Bight, 195
    Southern Bight, 174, 195
North Uist, 110
Norway, 135
    fish farming, 174–5
Norwegian Sea, 3, 8, 131–2, 134–5, 172, 176
notochord, 151

oceanography, 77, 91
oceans
    'buffering' in, 125
    circulation of currents, 77, 83, 107, 176
    warming of, 175–6
On the Origin of Species
    (Darwin), 81
Oncorhynchus genus, 196–7
Osborne House, 27
otters, 49, 202
Outer Hebrides, 132
ovarian fluid, 12
Owen, Sir Richard, 33

oxidation, 106, 109
oxygen, 167–8, 197–8

Pacific Ocean, 77, 90, 122, 127, 196–7
Palestine, 137
pandas, 160
Paris, 28, 116
parrs, 44–5, 48–9, 51, 65–6, 75, 174, 193, 201–5
parr markings, 48–9, 65–6, 205
and reproduction, 14–16, 118, 191
Parsons, Mr Midshipman, 31
Pastrana, Julia, 27–8
pattern recognition, 149
pearlsides, 95
Peel, Sir Robert, 111
perch, 146
Peripatus, 88
Perthshire, 193
Peterculter, 55
pheasants, 55
phosphates, 16, 106
photosynthesis, 106, 167–8
Picts, 22–3
Piggins, Dr David, 50
pike, 125, 146, 202
pipefish, 96
Pitlochry, Freshwater Fisheries Laboratory, 44, 144, 187
pituitary hormones, 135
plaice, 172
platypus, 87
poaching, 44
pollution, 23, 40, 43, 124
polyploid plants, 162
Polystichum fern, 27
porpoises, 171
Portsmouth, 78, 84, 86, 90
Presbyterianism, 139
puffins, 171

Quartenary Period, 173
quinnat (chinook) salmon, 122

rabbits, 31, 44
rainforests, 172
Ramón-y-Cajal, Santiago, 70–2
rats, 28–31, 44, 113, 142
recapitulation, theory of, 155
red mullet, 118
redds, 12, 14–16, 46, 50, 118, 145, 189, 191
Reformation, 137–8
reindeer, 116
Reliquiae Deluvianiae (William Buckland), 33
Rhine, River, 24
Roaring Forties, 78
Romans, 22, 118
Ross, Andrew, 57–8
Ross, Archibald, 56–62
Ross, Mary (later Adam), 57
Royal Navy, 30, 78–9, 83, 90, 93
Royal Norwegian research ship *Johan Hjort*, 133, 178
Royal Society, 81, 83
Royal Society of Edinburgh, 118
Ruskin, John, 35
Russia, 28

St Columba, 22
salmon
artificial culture, 117–26, 165, 174
central nervous system, 76
colouring, 65–6
direction-finding, 131
diving behaviour, 21
eating qualities, 109–10, 126
evolutionary history, 158–65, 195–9

fat reserves, 108–9, 128, 160, 176
feeding behaviour, 108–10
flesh colour, 108
homing behaviour, 50, 75–6, 120, 128, 135, 164, 193
and hybridization, 195
and ice ages, 177–8
kidneys, 64
lateral line, 4, 21, 131, 135
learning ability, 198
life cycle, 46–51, 157, 196
migrations, 75, 94–103, 120, 122, 130–6, 165, 178, 190, 193
mortality rates, 51, 129
nares, 5
paired fins, 76, 158
pectoral fins, 49
pelvic fins, 22
responsiveness to lures, 36
scales, 126–7
scarcity, 40
sense of smell, 37, 67, 75–6
sexual maturation, 4, 128, 147, 175
and siblings, 49
smolting process, 147
swim bladder, 3, 158, 197
water balance, 63–5
and water quality, 124–8
salmon, cock
and astaxanthin, 109
colouring, 5–6, 10, 12, 109
dorsal fin, 12, 22
early-run, 190–1
kype, 12–13, 190, 201
and reproduction, 12–16, 118
salmon, farmed, 44, 109–10, 172, 174–5
and genetic integrity, 175
and hybridization, 195

isolation and speciation, 193–5

salmon, hen
and astaxanthin, 109
and choice of spawning site, 50
colouring, 12, 15
early-run, 189–91
escaped, 195

salmon gaffs, 8

salmon populations
age structure and run-timing, 193
in Britain, 23–4
density-dependent mortality, 51
egg survival rates, 50–1
and fresh water variations, 97
locally distinct populations, 120, 165
North American populations, 110, 122, 193
refugial groups, 177–8

salmon trout, 195–6

sand eels, 4–5, 65, 94, 98, 130

Sarsia, 152

Saul, King, 138

scent marking, 49

Scotland
Bronze Age burials, 141
fish farming, 174–5
fishing rights, 23, 124
kelt rehabilitation, 36
Protestant Reformation, 137–9
salmon populations, 23–4

Scots, 22

Scotsman, The, 195

Scott, Lord John, 60

sea birds, diving, 198

sea horses, 96

sea lilies (crinoids), 87

Sea of Galilee, 20

sea trout, 96, 101–3, 105, 144
eating qualities, 110
feeding behaviour, 107–8, 110
hen fish, 146
and hybridization, 195
life cycle, 157, 195
and migration, 122, 144–7
progeny, 144–5
and water quality, 124–5
see also trout

seals, 5, 15, 61

Seceder Kirk, 139

Second Life Guards, 28

sequential olfactory imprinting, 75

Severn, River, 24

sewage treatment, 124

sex hormones, 14, 189

sexual selection, 185–6

shags, 95

sharks
basking, 173
six-gilled, 87

Shaw, John, 118

sheep, 70, 74

Sheerness, 79, 84

Shetland Islands, 4

shoals, 126–7

short kists, 141

Shuttleworth, Rev. Philip, 33

Sidney Sussex College, Cambridge, 113

silicon, 106

Simpson, Alex, 94

Simpson, Dr Tommy, 67, 69–70, 74, 76

Sir William Hardy, 69

Slessor, George, 99

slugs, 112

smelts, 160, 163–4, 197

Smith, Sir Sidney, 30–1

smolts, 4, 14–15, 45, 50–1, 65–7, 77, 120, 205
artificial culture, 125–6, 174
feeding behaviour, 97–8
fins and tail, 66
migrations, 94–103, 130–6
'post-smolts', 51, 98
shape, 66

snails, 35, 112

Société Impériale d'Acclimatation, 116

Society of Arts, 116

solar radiation, 168–9

Southern Ocean, 88

Spain, 120

sparling, 163

spermatozoa, 13–14, 118, 191

Spey, River, 44, 99, 101

spiders, 112

spring plankton bloom, 65

squirrels, 28, 31

steelhead trout, 196

sticklebacks, 43, 96–8

Stoker, Bram, 101

stone loach, 153

stoneflies, 47, 98

Stonehaven, 6

sulphur dioxide, 168

sulphuric acid, 124

Sutors, the, 94, 98

Svalbard, 180

'taking times', 36

Tasmania, 122

Tay, River, 145, 192

tectonic plates, 90

teeth, 76

teleosts, 157–8
and adipose dorsal fin, 160–1

Thames, River, 23–4, 43, 124, 153

Thomson, Captain Frank Thurle, 88

thyroid gland, 156
thyroid hormones, 75
Todd, Professor Chris, 176
Tomasson, Tumi, 126
Torres Straits, 91
Torry Research Station, 69
Trafalgar, Battle of, 93
trout, 43, 47, 49, 96, 191, 193
    artificial culture, 117–18,
        122–3
    cock fish, 145–6
    evolutionary history, 163–4,
        195
    hen fish, 107, 145–6
    and hybridization, 195–6
    and migration, 144–7
    sexual maturation, 147
    and Ythan catchment,
        140–7
    see also sea trout
tuna, blue fin, 173, 198
Tweed, River, 124, 171, 195
Tyne, River, 41, 124

United States of America, 41
University of Edinburgh,
    81, 83
University of Liverpool, 118
University of St Andrews,
    149, 176

urine, 64
Usher, Bishop, 33

vertebrates, 67, 70–2, 156,
    197–8
Victoria, Queen, 27, 79, 187
volcanic activity, 168–9
Voyage of the Beagle, The
    (Darwin), 80

Wales, 24
Walker, Andy, 144–5, 147
Walpole, Spencer, 35–6,
    38–9, 41
wapiti, 116
Wardle, Dr Clem, 160
water, 63–4
Waterloo, Battle of, 31
Wear, River, 41
Weil's disease, 142
Wellington, Duke of, 30
West Greenland, 172, 175,
    190
whales, 107, 178–85, 197–8
    Atlantic grey whale, 183,
        198
    brains of, 178, 183
    humpback whales, 179–80
    killer whales, 3, 180
    locomotion in, 198

migrations, 198
    minke whales, 179–80,
        183–4
    reproductive rates, 198
    sperm whales, 178, 180
White, Thomas, 123
whiting, 171–2
Why not eat insects? (Holt),
    111–13
Wilhelm II, Kaiser, 27
Wilkinson, Jonny, 71
Winchester College, 33, 35,
    38–9, 118
woodlice pills, 112
Woods, Tiger, 71
Wyville Thomson, Charles,
    81–4, 86–7, 90

yaks, 116
yolk sac, 46–8
Youngson, Alan, 46
Ythan, River, 141, 143, 147

Zoological Society of
    London, 114
zooplankton, 77, 106, 108,
    130
Zulu wars, 27